Future of Business and Finance

The Future of Business and Finance book series features professional works aimed at defining, analyzing, and charting the future trends in these fields. The focus is mainly on strategic directions, technological advances, challenges and solutions which may affect the way we do business tomorrow, including the future of sustainability and governance practices. Mainly written by practitioners, consultants and academic thinkers, the books are intended to spark and inform further discussions and developments.

Ken Huang • Carlo Parisi • Lisa JY Tan •
Winston Ma • Zhijun William Zhang
Editors

Web3 Applications Security and New Security Landscape

Theories and Practices

 Springer

Editors
Ken Huang (iD)
DistributedApps LLC
Fairfax, Virginia, USA

Carlo Parisi
Hacken
Lisbon, Portugal

Lisa JY Tan
Economics Design
Singapore, Singapore

Winston Ma
CloudTree Ventures
New York, USA

Zhijun William Zhang
Bank for International Settlements
Innovation Hub
Stockholm, Sweden

ISSN 2662-2467   ISSN 2662-2475   (electronic)
Future of Business and Finance
ISBN 978-3-031-58001-7   ISBN 978-3-031-58002-4   (eBook)
https://doi.org/10.1007/978-3-031-58002-4

This book is dedicated to the intrepid explorers of the Web3 ecosystem and the architects of the new security landscape, whose innovative spirit and unwavering dedication are shaping the future of digital technology and cybersecurity.

To the developers, cybersecurity experts, and enthusiasts, your commitment to unraveling the complexities of the Web3 space is inspiring. Your efforts in securing decentralized finance, non-fungible tokens, and decentralized autonomous organizations are pioneering a new era of digital security.

We honor the inquisitive minds delving into the realms of crypto asset exchange security and central bank digital currencies, whose pursuit of knowledge is crucial for navigating the intricacies of the decentralized world. Your curiosity and dedication are foundational to advancing secure and equitable digital interactions.

This book pays tribute to those addressing the challenges highlighted in its chapters: the intricate dance between Web3 and ransomware attacks; the multifaceted risks and innovations in the supply chain; the groundbreaking intersection of artificial intelligence with decentralized networks; the foresight required to anticipate and mitigate quantum attacks; and the vital importance of privacy-preserving computation in a digital era that values both transparency and confidentiality.

To our families, friends, and mentors, your unwavering support has been a beacon of light in this journey. Your belief in our vision has been instrumental in our quest to demystify the complex security landscape of Web3 and its interplay with emerging technologies.

"Web3 Application Security and New Security Landscape" is a testament to the collective wisdom, expertise, and forward-thinking of all those contributing to the security and integrity of the decentralized digital world. This book stands as a comprehensive guide and a tribute to the collaborative effort in navigating and securing the ever-evolving Web3 ecosystem.

Foreword 1

As the CEO of SkyBridge, a leading global investment firm, I have witnessed first-hand the transformative power of technology in the world of finance and beyond. In an era defined by innovation and digital disruption, security stands as a fundamental pillar upon which trust and progress are built.

"Web3 Application Security and New Security Landscape: Theories and Practices" is a remarkable addition to the discourse surrounding the security challenges and solutions in the Web3 era. Authored by experts who have immersed themselves in the intricate web of Web3 technologies, this book offers invaluable insights into safeguarding digital assets and ensuring the integrity of decentralized applications.

The digital landscape is evolving at an unprecedented pace, and Web3 technologies, including blockchain, decentralized finance (DeFi), non-fungible tokens (NFTs), and decentralized autonomous organizations (DAOs), are at the forefront of this revolution. These innovations hold immense potential, but they also introduce complex security considerations.

In my role as the CEO of SkyBridge, I understand the critical importance of staying ahead of the curve when it comes to security. This book provides a comprehensive overview of the theoretical foundations and practical strategies required to navigate the Web3 security landscape effectively. It covers a wide range of topics, from DeFi security and NFT protection to the challenges posed by quantum computing and the imperative of privacy-preserving computation.

Furthermore, I am pleased to highlight that this book is a companion to the previously published "A Comprehensive Guide for Web3 Security: From Technology, Economic, and Legal Aspects." Together, these two volumes offer a holistic perspective on Web3 security, addressing not only the technical aspects but also the economic and legal dimensions, making them essential references for anyone involved in the Web3 ecosystem.

In the dynamic world of finance and investment, where digital assets and decentralized platforms are becoming increasingly prevalent, security is not an option—it's a necessity. "Web3 Application Security and New Security Landscape: Theories and Practices" equips readers with the knowledge and strategies to navigate this exciting yet challenging landscape securely. I commend the authors for their dedication to advancing the field of Web3 security, and I am confident that this book will serve as a vital resource for professionals and decision-makers alike.

As we continue to explore new frontiers in the financial industry and beyond, security remains central to our mission of providing innovative and secure investment solutions. This book contributes significantly to the broader conversation about securing digital assets and fostering trust in the Web3 era. I encourage all those who are passionate about the future of technology and finance to engage with this comprehensive and insightful resource.

New York, NY, USA   Anthony Scaramucci

Foreword 2

As a Vice President and Professor at the Hong Kong University of Science and Technology (HKUST) and Chairman of the Academic Executive Committee at Institute of Web3.0 Hong Kong, I have had the privilege of witnessing the remarkable fusion of technology, education, and research that defines our institution. In this era of profound technological transformation, one of the most pressing challenges we face is ensuring the security and integrity of our digital infrastructure and applications.

"Web3 Application Security and New Security Landscape: Theories and Practices" is an essential contribution to the ongoing discourse surrounding security in the Web3 era. Authored by experts deeply immersed in the intricate Web3 ecosystem, this book provides valuable insights into securing decentralized applications and digital assets, making it a vital resource for professionals, researchers, and educators.

The digital landscape is evolving at an unprecedented pace, and Web3 technologies, such as blockchain, decentralized finance (DeFi), non-fungible tokens (NFTs), and decentralized autonomous organizations (DAOs), represent the vanguard of this transformation. These innovations hold the promise of democratizing access to information and resources, but they also introduce new and complex security challenges.

In my capacity at HKUST, I appreciate the significance of staying at the forefront of technological advancements, particularly in the realm of security. This book offers an expansive exploration of the theoretical foundations and practical strategies required to navigate the Web3 security landscape effectively. It covers an array of topics, from DeFi security and NFT protection to quantum computing threats and privacy-preserving computation techniques.

I am pleased to note that this book is a companion to the previously published "A Comprehensive Guide for Web3 Security: From Technology, Economic, and Legal Aspects." Together, these two volumes provide a comprehensive view of Web3 security, addressing not only the technical dimensions but also the economic and legal facets, rendering them indispensable references for those engaged in the Web3 ecosystem.

In the dynamic field of technology and academia, where innovation and research intersect, security remains a cornerstone. "Web3 Application Security and New Security Landscape: Theories and Practices" equips readers with the knowledge and tools to navigate this transformative landscape securely. I commend the authors for their dedication to advancing the field of Web3 security, and I am confident that this book will serve as a critical resource for professionals, scholars, and students.

As HKUST continues its commitment to pioneering research and education at the intersection of technology and society, we recognize the pivotal role that security plays in shaping our digital future. This book contributes significantly to the ongoing dialogue about securing digital assets and advancing knowledge in the Web3 era. I encourage all those who are passionate about technology, education, and research to engage with this insightful and comprehensive resource.

Pak Shek Kok, Hong Kong, PR China   Yang Wang
Clear Water Bay, Hong Kong, PR China

Foreword 3

I am deeply honored to pen this foreword for "Web3 Application Security and New Security Landscape: Theories and Practices" authored by a distinguished group of experts in the field. This book is a testament to their commitment to advancing our understanding of the intricate security challenges within the Web3 landscape.

In today's digital age, where Web3 technologies are reshaping industries and economies, security has taken center stage. As the Executive Chairman of the World Digital Technology Academy (WDTA), Chairman of Cloud Security Alliance Greater China (CSA GCR), and a Foreign Academician of the Ukrainian Academy of Engineering Sciences, I have had the privilege of witnessing the rapid evolution of technology and its profound impact on our world.

The chapters within this volume represent a comprehensive exploration of Web3 security, spanning from DeFi and NFTs to DAOs, crypto asset exchanges, CBDCs, and beyond. Each chapter provides a deep dive into the security intricacies specific to these domains, offering real-world case studies and practical insights that illuminate the path toward robust security practices.

In conjunction with the evolving Web3 landscape, this volume meticulously addresses the emerging threat landscape. Chapters spanning from Web3 and ransomware attacks to quantum threats and privacy-preserving computation shed light on the novel challenges that confront our digital future. It provides guidance on tackling these challenges head-on, ensuring the security and integrity of Web3 applications.

As we navigate the complexities of the Web3 era, it is paramount that we proactively address security challenges. The authors of this volume not only diagnose the security landscape but also prescribe defense measures and best practices that empower organizations and individuals to secure their digital assets and contribute to the integrity of the Web3 ecosystem.

This volume is more than an academic endeavor; it is a pragmatic guide for those seeking to navigate the intricate security terrain of Web3 applications. It calls upon readers to engage actively with the security challenges and opportunities that define this transformative era.

I commend the authors for their dedication to advancing the discourse on Web3 security, with a keen focus on the new and evolving threat landscape. Their expertise and commitment shine through in every chapter, making this volume an indispensable resource for security professionals, developers, policymakers, and anyone navigating the ever-changing security landscape of Web3 applications.

As you embark on this enlightening journey through the security theories and practices within the Web3 landscape, I encourage you to absorb the insights offered within these pages. May this volume serve as your trusted companion in securing the digital future.

Geneva, Switzerland   Yale Li
Shanghai, China
Kyiv, Ukraine

Foreword 4

As the Chief Information Security Officer (CISO) of the World Bank, I have the privilege of overseeing the security of a vast and complex digital ecosystem that spans the globe. In an era defined by rapid technological advancements and digital transformation, ensuring the security and integrity of our digital assets is paramount.

"Web3 Application Security and New Security Landscape: Theories and Practices" arrives at a crucial juncture in our digital journey. It is a profound testament to the evolving nature of the digital landscape and the critical role that security plays in the Web3 era. This book serves as an invaluable resource for security professionals, technologists, and decision-makers alike, offering a comprehensive understanding of the security challenges and opportunities presented by Web3 technologies.

In my role at the World Bank, I understand the importance of staying ahead of the curve when it comes to security. The emergence of Web3, with its decentralized applications, blockchain networks, and novel digital assets, presents both exciting possibilities and complex security implications. This book, authored by experts who have dedicated themselves to unraveling the intricacies of Web3 security, provides deep insights into the theoretical foundations and practical strategies needed to protect Web3 applications and systems.

The book's coverage of decentralized finance (DeFi), Non-Fungible Tokens (NFTs), Decentralized Autonomous Organizations (DAOs), crypto asset exchanges, Central Bank Digital Currencies (CBDCs), and quantum threats reflects the diverse and dynamic nature of the Web3 security landscape. It not only identifies emerging threats but also offers actionable solutions and best practices, making it an indispensable guide for safeguarding digital assets and infrastructures in this new era.

Moreover, I must emphasize that this book is a companion to the previously published "A Comprehensive Guide for Web3 Security: From Technology, Economic, and Legal Aspects." Together, these two volumes offer a holistic perspective on Web3 security, encompassing technological, economic, and legal dimensions. They are, without a doubt, must-read references for anyone seeking to secure the Web3 ecosystem comprehensively.

In the ever-evolving realm of cybersecurity, knowledge is the most potent weapon. "Web3 Application Security and New Security Landscape: Theories and Practices" equips readers with the knowledge and insights necessary to navigate the intricate and transformative landscape of Web3 securely. I commend the authors for their dedication to advancing the field of Web3 security, and I encourage all those who are responsible for safeguarding digital assets to engage with this invaluable resource.

As the World Bank continues to embrace digital innovation in our mission to alleviate poverty and promote sustainable development worldwide, I recognize the importance of robust security measures. This book is a vital contribution to the global dialogue on digital security and resilience, and I am confident it will empower and inspire the next generation of security professionals and technologists to protect the Web3 ecosystem effectively.

Washington, DC, USA   Clay Lin

Foreword 5

With a career spanning two decades across various industries, I've had the distinct pleasure of working alongside some of the most forward-thinking minds in the realm of emerging technologies. Our mission has been to weave cutting-edge innovations such as Blockchain, Generative AI, AR/VR, and Quantum Computing into the fabric of our strategic initiatives. This journey has allowed us to not only chart the course for innovation within the banking sector but also to cultivate an ecosystem ripe for technological breakthroughs. Central to this mission has been our emphasis on collaboration with researchers, innovators, and fintech firms, fostering a culture where the exchange of ideas propels us forward.

Among those I've had the privilege to collaborate with is Ken Huang, a luminary in the fields of Generative AI and Web3. Our joint endeavors, spanning various book projects and thought leadership efforts, have underscored the value of interdisciplinary collaboration in navigating the complexities of today's digital landscape. It is within this context of mutual respect and shared vision that I find myself contributing the foreword to "Web3 Application Security and New Security Landscape." Under the stewardship of Ken Huang as chief editor and the collective wisdom of an esteemed team of co-editors, this book emerges as a seminal piece of literature in the domain of Web3 security.

The book meticulously dissects the multifaceted security challenges that beset the Web3 ecosystem, offering readers a comprehensive exploration of topics ranging from DeFi, NFT, and DAO security, to the intricacies of Crypto Asset Exchange and CBDC security. Each chapter, authored by experts who bring a wealth of knowledge and insight, not only shines a light on the vulnerabilities that have surfaced with the advent of Web3 technologies but also charts a course toward resilience and robust defense mechanisms.

What sets this work apart is its forward-looking perspective, particularly evident in the discussion of frontier security concerns. This section delves into the confluence of Web3 with technologies like AI and Quantum Computing, articulating potential vulnerabilities while proposing innovative safeguards. It's a testament to the book's depth and breadth that it not only navigates current security paradigms but also ventures into speculative territory, preparing readers for the emerging threats and opportunities on the horizon.

Reflecting on our past collaborations, it's evident that the synergy between Ken Huang and myself, along with our collective experiences in Generative AI and Web3, has enriched the discourse on digital security and innovation. This book, therefore, stands not just as a compendium of expert knowledge, but as a beacon for those committed to forging secure, inclusive, and innovative digital futures.

For developers, strategists, policymakers, and anyone vested in the evolution of Web3 technologies, "Web3 Application Security and New Security Landscape" offers indispensable insights. As we continue to venture into new frontiers of digital interaction and finance, the guidance of Ken Huang and the distinguished team of co-editors provides a compass by which we can navigate the complexities of this evolving digital ecosystem with clarity and confidence.

Charlotte, NC, USA   Jyoti Ponnapalli

Foreword 6

I am delighted to provide this foreword for "Web3 Application Security and New Security Landscape: Theories and Practices," a groundbreaking work that arrives at a moment of profound significance in the realms of technology, finance, and security. As the Chairman of the Board at the Global FinTech Institute and Co-Chair of the Blockchain Security Alliance, in addition to my role as a Professor at the Singapore University of Social Sciences (SUSS), I have devoted my career to advancing the understanding of financial technology and blockchain security. This book stands as a remarkable contribution to our collective mission.

The world is in the throes of a digital revolution that is reshaping industries, economies, and societies. Web3 technologies, encompassing blockchain, smart contracts, decentralized applications, and even Generative AI, are at the vanguard of this transformation. They hold the promise to redefine how we conduct business, govern, and interact online. However, with the expanding scope and complexity of these technologies comes an array of intricate security challenges.

This book, meticulously crafted by experts who are luminaries in their respective domains, delves deep into the multifaceted landscape of Web3 security. It commences with an exploration of DeFi security, illuminating the exponential growth of the decentralized finance ecosystem and the paramount significance of security within it. From NFTs to DAOs, crypto asset exchanges to CBDCs, quantum threats to privacy-preserving computation, and even the intersection of Generative AI and Web3 security, the book offers comprehensive coverage of the security dimensions that define the Web3 era.

Importantly, this book is more than an academic exercise; it is a pragmatic guide that furnishes readers with real-world case studies, best practices, and actionable recommendations. It equips individuals with the knowledge and tools indispensable for navigating the intricate security landscape of Web3 effectively.

It is worth noting that this book is also a companion to the previously published "A Comprehensive Guide for Web3 Security: From Technology, Economic, and Legal Aspects." Together, these two volumes provide a holistic view of Web3 security, spanning technological, economic, legal, and even generative AI dimensions.

They serve as an essential resource for those endeavoring to secure the Web3 ecosystem comprehensively.

In my capacities at the Global FinTech Institute and the Blockchain Security Alliance, I have borne witness to the critical importance of collaboration, education, and research in advancing the security of emerging technologies. "Web3 Application Security and New Security Landscape: Theories and Practices" encapsulates these principles, fostering a deeper understanding of the security challenges and opportunities that define Web3 technologies, including Generative AI.

Standing at the threshold of a new era marked by decentralization, trustlessness, and boundless innovation, security remains the linchpin that fortifies the promise of Web3. I commend the authors for their unflagging dedication to advancing the field of Web3 security, and I am confident that this book will stand as an indispensable resource for professionals, researchers, educators, and decision-makers alike.

In closing, I extend my warm congratulations to the authors for their exceptional work in compiling this comprehensive and insightful resource. I encourage all who are passionate about the future of technology, finance, and security, including the nuanced intersection of Generative AI and Web3 security, to engage wholeheartedly with this profound contribution to the discourse.

Singapore, Singapore   David Lee Kuo Chuen

Preface

In the evolving landscape of Web3 security, we are at the dawn of a new era, where the complexities of today's threats meet the pioneering spirit of tomorrow's defenders. This journey is not merely about confronting challenges but about seizing the opportunity to redefine what security means in a decentralized world. As we explore the intricate interplay between cutting-edge technologies and emerging security threats, we are reminded of our capacity for innovation, resilience, and the relentless pursuit of a secure, decentralized future. Each obstacle encountered on this path reinforces our resolve, encouraging us to think more creatively and build more robust defenses, laying the groundwork for a digital world that is not only safer but also inherently more free and equitable for generations to come.

Our book, "Web3 Application Security and the New Security Landscape," is an in-depth examination of the evolving challenges and cutting-edge solutions in the security of the Web3 ecosystem. Designed for a broad audience, this work is intended for developers, cybersecurity experts, and enthusiasts intrigued by the confluence of blockchain technology, decentralized applications (dApps), and the security of digital assets. As we navigate the shift toward a decentralized digital landscape, it becomes imperative to reassess conventional cybersecurity methodologies. The inherently decentralized nature of Web3 technologies calls for innovative and dynamic security strategies, tailored to address the unique risks and opportunities these technologies present. This book is structured to demystify such strategies, providing a comprehensive exploration into the security aspects specific to the Web3 domain.

The narrative of the book is divided into three distinct parts. Part I encompasses Chaps. 1–5, delving into the security of Web3 applications. It starts with an examination of "DeFi Security," highlighting the unique challenges and vulnerabilities present within decentralized finance. This is followed by "NFT Security," which investigates the risks and protective measures linked to non-fungible tokens. "DAO Security" examines the security considerations within decentralized autonomous organizations, introducing readers to the concepts of digital governance and operations. The discussion extends to "Crypto Asset Exchange Security," emphasizing the importance of securing cryptocurrency exchanges, and wraps up with "CBDC

Security," offering insights into the security dimensions of central bank digital currencies.

Part II, running from Chaps. 6–10, focuses on the frontier aspects of Web3 security. It begins with "Web3 and Ransomware Attacks," illustrating the changing landscape of cyber threats in a decentralized environment. "Web3 and Supply Chain Risks" evaluates how blockchain technology can enhance transparency and security within supply chains. "Web3 and AI Security" explores the intersection between generative AI and decentralized networks, assessing the potential benefits and pitfalls. "Web3 and Quantum Attacks" addresses the imminent threats quantum computing poses to the cryptographic underpinnings of Web3. The section concludes with "Privacy-Preserving Computation and Web3," discussing strategies for upholding privacy in an increasingly digital world.

Part III consists of Chap. 11, "Summary and Future Trends," which synthesizes the insights and learnings from the preceding chapters, offering a holistic view and prospective outlook on the future of Web3 security.

This book aims to serve as an essential resource for those involved in the development of Web3 applications and for readers keen on expanding their understanding of this field. As Web3 technology progresses, the imperative for heightened security consciousness and proactive measures becomes increasingly pronounced. Through "Web3 Application Security and the New Security Landscape," we aspire to equip our readers for the challenges and opportunities that lie ahead.

Fairfax, Virginia, USA   Ken Huang

Acknowledgment

As the chief editor of "Web3 Applications Security and New Security Landscape: Theories and Practices," I extend my heartfelt gratitude to my fellow editors, Carlo Parisi, Lisa JY Tan, William Zhang, and Winston Ma for their unwavering dedication and expertise that have significantly shaped this publication. Their profound knowledge and insightful perspectives have been instrumental in guiding the development of this comprehensive work, ensuring that it not only explores the intricacies of Web3 security in depth but also remains relevant and accessible to a diverse readership.

The collaborative efforts of an exceptional group of authors have been essential in creating a resource that addresses a wide array of topics within the sphere of Web3 security. The depth and diversity of their expertise, covering everything from decentralized finance and non-fungible token security to the emerging challenges posed by quantum computing and central bank digital currencies, provide readers with a nuanced understanding of the current and future security landscape in the Web3 domain.

I would like to express special appreciation to contributors such as Jerry Huang, whose extensive experience in security, artificial intelligence, and machine learning significantly enhances the content of this book. The insights from Krystal Jackson, Sean Heide, Jennifer Toren, Dr. Luyao Zhang, and Mudi Xu, each an authority in their respective area, add a comprehensive and multifaceted perspective to our discussion of Web3 security. Their contributions span from practical cybersecurity applications to theoretical advancements in blockchain, cloud security, and AI, enriching the book with both depth and breadth.

I would also extend my deepest gratitude to Anthony Scaramucci, CEO of SkyBridge and former White House Director of Communications; Prof. Yang Wang, Vice President and Professor at the Hong Kong University of Science and Technology (HKUST) and Chairman of the Academic Executive Committee at the Institute of Web3.0 Hong Kong; Dr. Yale Li, Executive Chairman of the World Digital Technology Academy (WDTA), Chairman of the Cloud Security Alliance Greater China (CSA GCR), and Foreign Academician of the Ukrainian Academy of Engineering Sciences; Clay Lin, Chief Information Security Officer (CISO) of the

World Bank; Jyoti Ponnapalli, SVP of Innovation Strategy and Research, Truist; and David LEE Kuo Chuen, Chairman of the Board at the Global FinTech Institute, Co-Chair of the Blockchain Security Alliance, and Professor at the Singapore University of Social Sciences (SUSS), for their invaluable contributions by writing the forewords for this book. Their forewords not only enrich the discourse on the security challenges and advancements in the Web3 era but also provide readers with unique insights from their distinguished careers. Their collective expertise across the fields of finance, academia, technology, and security profoundly reflects this book's depth and relevance.

Additionally, I express my sincere appreciation to Professor Feng Zhu from Harvard Business School, Professor Xi Chen from New York University, Dr. Yao Qian who was Ex-Head of Chinese CBDC Program, Dr. Youwei Yang who is Chief Economist of a NYSE-listed Bitcoin mining company, and Ms. Grace Huang, Project Manager of PIMCO, for their strong endorsements. Their recommendations underscore the significance of this work in the realms of academia, research, and industry. Their endorsements not only validate the book's comprehensive approach to Web3 security but also highlight its practical utility for professionals, students, and researchers alike.

The collaboration and support of these distinguished individuals underscore the critical importance of security in the evolving digital age, providing readers with the knowledge and tools necessary to secure the digital future. Their contributions, coupled with their encouragement and insights, play a pivotal role in advancing the understanding and awareness of security in the Web3 landscape via this publication.

"I wholeheartedly recommend 'Web3 Application Security and New Security Landscape: Theories and Practices' as a must-read for anyone interested in the evolving realm of Web3 security. The authors have skillfully compiled a wealth of knowledge that is not only academically rigorous but also deeply practical. This book will serve as an indispensable resource for students, researchers, and professionals seeking to understand and navigate the complexities of Web3 security."

—Professor Feng Zhu, Harvard Business School

"As an academician in the field of AI, Web3, Blockchain and Distributed Ledger Technologies, I find 'Web3 Application Security and New Security Landscape: Theories and Practices' to be a comprehensive and insightful resource. The book's coverage of diverse topics, from DeFi to quantum threats, is commendable. It not only sheds light on the theoretical foundations but also provides practical solutions. I recommend this book to my students and colleagues as an essential guide to understanding the security challenges and opportunities in the Web3 era."

—Professor Xi Chen, NYU Stern School of Business

"This book is essential for anyone looking to understand and implement the highest standards of digital security in the context of CBDCs and beyond. Its concise yet comprehensive coverage makes it an invaluable asset for professionals at the forefront of digital currency innovation.

Yao Qian, the first director of the Chinese Central Bank's Digital Currency (CBDC) Program and now Director of the Science and Technology Supervision Bureau of the China Securities Regulatory Commission."

—Yao Qian, Ex-Head of Chinese CBDC Program

"As the Chief Economist of a Bitcoin mining company listed on NYSE, I appreciate the significance of security in the blockchain and cryptocurrency space. 'Web3 Application Security and New Security Landscape: Theories and Practices' is a critical resource that addresses the security challenges and opportunities in Web3 technologies, which are closely intertwined with the cryptocurrency ecosystem. The book offers valuable insights into safeguarding digital assets and decentralized applications, making it essential reading for professionals and investors in the blockchain industry."

—Dr. Youwei Yang, Chief Economist at NYSE listed Mining Corp.

"As a Product Manager at PIMCO, where we continuously seek to integrate cutting-edge technology and security practices into our investment management processes, I find the second volume on Web3 security to be an exceptional resource. This book provides a comprehensive exploration of the security challenges and opportunities presented by decentralized technologies, which are increasingly relevant to our industry.

The practical insights and forward-looking analysis offered by the authors make this publication a must-read for professionals looking to navigate the complexities of Web3 with confidence and expertise. It adeptly addresses the nuanced demands of digital security in the financial sector, making it an invaluable tool for anyone involved in the intersection of finance and technology.

I highly recommend this book to colleagues and industry peers as a guide to mastering the intricacies of Web3 security and leveraging its potential to enhance our digital infrastructure."

—Ms. Grace Huang, Product Manager at PIMCO

Contents

About the Editors

Ken Huang is the author and chief editor of 8 books on Generative Artificial Intelligence and Web3, published, respectively, by international publishers including Springer, Cambridge University Press, John Wiley, and China Machine Press. He currently serves as the CEO of the AI and Web3 consulting and education company DistributedApps.AI, based in the United States. Additionally, he holds multiple roles including the expert member of the Blockchain Committee of the Chinese Institute of Electronics, the Co-Chair of AI Organization Responsibility Working Group at Cloud Security Alliance, and Chair of the Blockchain Security Working Group at the Cloud Security Alliance, GCR. He is also a core contributor to the Generative AI Working Group at the NIST and a core author of the OWASP Top 10 for LLM Applications.

Ken Huang has been invited to provide Speaking or Consulting services at institutions including the University of California, Berkeley, Stanford University, Peking University, Tsinghua University, Shanghai Jiao Tong University, China Pacific Insurance, and the World Bank in the past.

Moreover, he has given keynote speeches at international conferences, such as:

– The Davos World Economic Forum 2020 Blockchain Conference
– Consensus 2018 in New York
– The American ACM AI & Blockchain Decentralized Annual Conference 2019
– IEEE Technology and Engineering Management Society Annual Meeting 2019

- Silicon Valley World Digital Currency Forum
- Sino-US Blockchain Summit in Silicon Valley

He has also been awarded the "Blockchain 60" Figure Award by the National University Artificial Intelligence and Big Data Innovation Alliance Blockchain Special Committee in China in 2021.

Address: 12870 Williams Meadow Court, Fairfax, VA, 20171

Carlo Parisi   is a senior Solidity smart contract developer, senior auditor, and content creator with a degree in Computer Science. He has a deep knowledge in developing and auditing Solidity code, the main language used in Ethereum for smart contracts. He has been a Bitcoin Enthusiast since 2013, DeFi user since 2018.

Address: Hacken, Via Goffredo di Crollalanza n°2, 70121 Bari, Italy

Lisa JY Tan   work as the founder and lead economist at Economics Design has made her a pioneer in the design and engineering of digital ecosystems. With a track record of over 30 token economies and 50 token analyses, Lisa's work is characterized by a research-focused approach and a deep understanding of the potential of blockchain technology. As a highly sought-after speaker at conferences and forums worldwide, Lisa's expertise in token economics and DeFi has established her as a respected authority in the field of digital ecosystems.

Lisa JY Tan, Economics Design, Robinson Road 36, 068877 Singapore, Singapore

Zhijun William Zhang is the Technology and Innovation Adviser at the Bank for International Settlements (BIS) Innovation Hub—Nordic Centre, where he focuses on cybersecurity and resilience for future financial market infrastructure. Before joining the BIS, he was the lead security architect at the World Bank Group, where his team was responsible for security architecture design and assessment of all technology platforms and business solutions. He also led the security and risk work for WBG's innovation lab. Before joining the WBG, William worked at The

Vanguard Group in various capacities, including user experience design, emerging technology research, system architecture, and information security. William received his BS degree from Peking University, and his Ph.D. from the University of Maryland, both in computer science.

Address: Bank for International Settlements Innovation Hub, MÄSTER SAMUELSGATAN 36, 111 57, Stockholm, Sweden

Winston Ma   CFA is an investor, author, and adjunct professor in the digital economy. He is one of a small number of native Chinese who has worked as investment professionals and practicing capital markets attorneys in both the United States and China. Most recently for 10 years, he was Managing Director and Head of the North America Office for China Investment Corporation (CIC), China's sovereign wealth fund.

Prior to that, Mr. Ma served as the deputy head of equity capital markets at Barclays Capital, a vice president at J.P. Morgan investment banking, and a corporate lawyer at Davis Polk & Wardwell LLP in New York.

At CIC's inception in 2007, he was among the first group of overseas hires by CIC, where he was a founding member of both CIC's Private Equity Department and later the Special Investment Department for direct investing (Head of CIC North America office 2014–2015). He had leadership roles in global investments involving financial services, technology (TMT), energy, and natural resources sectors, including the setup of West Summit (Huashan) Capital, a cross-border growth capital fund in Silicon Valley, which was CIC's first overseas tech investment. For global investments, he served on the board of internationally listed and private companies.

A nationally certified Software Programmer as early as 1994, Mr. Ma is the book author of China's Mobile Economy (Wiley 2016, among "best 2016 business books for CIOs"), Digital Economy 2.0 (2017 Chinese), The Digital Silk Road (2018 German), China's AI Big Bang (2019 Japanese), and Investing in China (Risk Books, 2006). His new books are "The Hunt for Unicorns: How Sovereign Funds Are Reshaping Investment in the Digital Economy" (Wiley October 2020) and "The Digital War—How China's Tech Power Shapes the Future of AI, Blockchain, and Cyberspace"

(Wiley January 2021). He was selected as a 2013 Young Global Leader at the World Economic Forum (WEF) and has been a member of the Council for Long-Term Investing and the Council for Digital Economy and Society. He has been a member of the New York University (NYU) President's Global Council since its inception, and in 2014 he received the NYU Distinguished Alumni Award.

Address: CloudTree Ventures, 217 E 96th St, New York, NY 10128-3993, USA

Part I
Web3 Applications Security

The emergence of decentralized systems and Web3 has opened new possibilities for innovation across industries. However, these innovations also introduce unique security risks that must be understood and mitigated.

This first part of the book offers an in-depth analysis into the security landscape across major applications of Web3 today—DeFi, NFTs, DAOs, crypto exchanges, and CBDCs.

Chapter 1 sets the stage with an examination of inherent risks in the rapidly expanding DeFi ecosystem. It summarizes top DeFi breaches to date and best practices around audits and secure smart contract development. Chapter 2 covers various threats surrounding NFTs—exchange hacks, social engineering scams, smart contract exploits, and risks from decentralized storage.

Decentralized autonomous organizations are the backbone enabling collaboration and governance in Web3. Chapter 3 analyzes high profile attacks on DAOs as well as challenges like proposal analyses and decentralized decision-making. It also offers defense strategies around access controls, formal verification of systems and processes.

For the average user, centralized exchanges are still the primary gateway to crypto assets and Web3 applications. Chapter 4 deep dives into techniques used by exchanges to prove solvency, importance of custody solutions and secrets management. Case studies offer learnings from both catastrophic exchange hacks and demonstrations of resilience.

Finally, Chap. 5 discusses the unique security considerations involved in designing central bank digital currencies to maintain integrity of monetary systems. It examines risks from financial crimes as well as preserving privacy in CBDC implementations.

Together, these five chapters provide a 360° view into the security fundamentals underpinning much of the innovation in Web3 today—across DeFi, NFTs, DAOs, exchanges, and CBDCs. A solid grasp of these Web3 applications security topics is imperative for engineers, developers, architects, and analysts working on decentralized systems.

Chapter 1
DeFi Security

Carlo Parisi and Dmitriy Budorin

Abstract This chapter provides a comprehensive overview of the rapidly evolving decentralized finance (DeFi) ecosystem, emphasizing the critical importance of security within this space.

DeFi has experienced remarkable growth in recent years, disrupting traditional finance systems. DeFi is defined, and its various components, including decentralized exchanges, lending platforms, and yield farming protocols, are explored. The (de)centralized nature of DeFi is also discussed, empowering users while presenting unique challenges.

Security takes center stage. Inherent risks and vulnerabilities in DeFi are described deeply, with a focus on the potential consequences of security breaches for users and the ecosystem. Real-world examples of DeFi hacks illustrate the impact of such incidents. Importance of security audits and best practices is stressed to address these risks, particularly concerning smart contracts that allow DeFi protocols to exist. Rigorous audits and adherence to industry standards can significantly reduce vulnerabilities.

1.1 Introduction

1.1.1 Brief Overview of the Rapid Growth of the DeFi Ecosystem

Finance encompasses various aspects related to money, including currency creation, management, and the study of capital assets. It can be broadly defined as the field that deals with monetary matters.

C. Parisi (✉) · D. Budorin
Hacken, Lisbon, Portugal
e-mail: c.parisi@hacken.io; d.budorin@hacken.io

© The Author(s), under exclusive license to Springer Nature
Switzerland AG 2024
K. Huang et al. (eds.), *Web3 Applications Security and New Security Landscape*,
Future of Business and Finance, https://doi.org/10.1007/978-3-031-58002-4_1

Decentralized Finance aims to decentralize financial tools and processes. Traditionally, these tools have been centralized and lack censorship resistance. DeFi endeavors to address this issue, although, as we'll explore further, it hasn't fully resolved centralization concerns in many projects.

The history of DeFi traces back to Bitcoin's emergence in 2009, which, while not considered traditional DeFi, introduced the concept of decentralized payments. However, Bitcoin's simplicity limited its capacity to support complex financial services such as lending and borrowing.

Ethereum, launched in 2015, revolutionized the DeFi landscape. It offered a smart contract platform with a versatile programming language, Solidity, among others, and the ERC20 standard for token creation. This made Ethereum a preferred platform for DeFi development.

Significant developments in DeFi began in 2017 but gained substantial adoption in 2020 during the "DeFi Summer." This period was initiated by Compound's COMP token liquidity mining program, sparking a surge in yield farming and liquidity mining across various projects. Yearn Finance, exemplified by the YFI token, epitomized this rapid growth.

Decentralized exchanges (DEXs) played a crucial role in sustaining DeFi Summer. DEXs facilitate the exchange of tokens, significantly increasing the velocity of money within the DeFi ecosystem.

1.1.2 The Importance of Security in the DeFi Space

Security is a crucial concern within the DeFi space, given that every transaction involves tokens, often with substantial value. In contrast to web2, where hackers face barriers to accessing vast sums of money directly, DeFi's architecture makes it a prime target for malicious actors. Several key security risks are central to the DeFi landscape.

First and foremost is the security of smart contracts. These are meticulously examined by auditing firms to identify vulnerabilities and ensure their robustness. Auditors play a critical role in mitigating potential risks.

Additionally, personal security is a vital aspect, particularly the protection of private keys held by project administrators. Despite the term "decentralized finance," many DeFi projects retain varying degrees of centralization. Certain addresses may possess unique privileges or actions they can execute on the platform. In cases where the private key of such individuals is compromised, the funds within the project could be vulnerable to theft.

This underscores the need for stringent security practices and robust measures to safeguard both smart contracts and private keys within the DeFi ecosystem.

1.2   Understanding DeFi

1.2.1   Definition and Scope of DeFi, Including Its Various Components

DeFi is a blockchain-based financial ecosystem that seeks to recreate and innovate traditional financial services and products in a decentralized, permissionless, and trustless manner. DeFi eliminates the need for intermediaries, such as banks and financial institutions, by leveraging smart contracts and blockchain technology to facilitate various financial activities.

The scope of DeFi encompasses a wide range of financial services and components, including:

Decentralized Exchanges: DeFi's scope includes decentralized exchanges like Uniswap, SushiSwap, and Curve Finance. These platforms enable users to trade cryptocurrencies directly from their wallets, eliminating the need for intermediaries. DEXs leverage smart contracts to facilitate peer-to-peer trading and provide liquidity.

Lending and Borrowing Protocols: DeFi offers a robust lending and borrowing ecosystem. Users can lend their assets to earn interest or borrow assets by collateralizing their holdings. Prominent examples include Aave, Compound, and MakerDAO: These protocols allow users to access liquidity without relying on traditional banks.

Stablecoins: Stablecoins are a fundamental component of DeFi. These cryptocurrencies aim to maintain a stable value, often pegged to fiat currencies like the US Dollar. DeFi hosts various stablecoin projects, such as DAI, USDC, and USDT. Stablecoins provide a stable medium of exchange within the volatile cryptocurrency market.

Yield Farming and Liquidity Provision: DeFi introduces yield farming and liquidity provision strategies. Users can optimize their returns by providing liquidity to decentralized platforms and earning rewards. Yield farming involves moving assets between different DeFi protocols to maximize yield, showcasing the innovation within the space.

Derivatives and Synthetic Assets: DeFi expands into derivatives and synthetic assets, allowing users to gain exposure to a wide range of financial instruments without owning the underlying assets. These platforms enable users to hedge their positions and speculate on market movements.

Oracles: Oracles play a crucial role in DeFi by providing external data to smart contracts. These data sources supply real-world information such as price feeds, market data, and event outcomes, enabling smart contracts to make informed decisions.

Decentralized Autonomous Organizations (DAOs): DAOs are community-driven organizations that employ token-based governance to make collective decisions. Many DeFi projects are governed by DAOs, enabling decentralized control and decision-making over protocol upgrades and changes.

Insurance: DeFi insurance protocols such as Nexus Mutual provide coverage against smart contract vulnerabilities and other risks. Users can purchase insurance to protect their assets in the event of unforeseen events.

Cross-Chain and Interoperability Solutions: DeFi explores cross-chain compatibility and interoperability to connect different blockchain networks. These initiatives enhance liquidity and functionality across the DeFi ecosystem, enabling assets to move seamlessly between blockchains.

Tokenization of Real-World Assets: DeFi aims to tokenize real-world assets like real estate, stocks, and commodities. This process involves representing these assets as digital tokens, making them tradable on blockchain platforms and enhancing their liquidity.

Decentralized Identity and Reputation Systems: Some DeFi projects focus on identity verification and reputation scoring to reduce risks associated with anonymous transactions. These systems contribute to the security and trustworthiness of DeFi interactions.

DeFi's scope is continually expanding as developers and entrepreneurs innovate and explore new possibilities within the decentralized financial landscape. It offers the potential to revolutionize how individuals access, manage, and interact with financial services, promoting financial inclusion and reducing reliance on traditional financial intermediaries.

1.2.2   The (De)Centralized Nature of DeFi Protocols and Applications

When we explore the landscape of DeFi, one immediately encounters the term "decentralization" as a defining feature. However, it is essential to recognize that decentralization is not an all-or-nothing concept but rather exists along a spectrum. DeFi, short for Decentralized Finance, implies that these financial systems and applications aim to operate without the central authority, such as traditional banks or intermediaries. Nevertheless, the degree of decentralization can significantly vary across different DeFi protocols and applications, leading us to question the appropriateness of the term "DeFi" itself in some cases.

One of the primary challenges to consider is that certain DeFi protocols and applications may not achieve a high level of decentralization, or in some instances, they may be inherently unable to achieve full decentralization. Take, for example, the tokenization of real-world assets. In this scenario, DeFi platforms seek to represent physical assets, like real estate or art, as digital tokens on a blockchain. However, the very nature of these real-world assets often necessitates the involvement of trusted custodians. These custodians play a critical role in certifying the legitimacy and ownership of the underlying assets, introducing a centralized element within the DeFi ecosystem.

Another aspect of DeFi that challenges its decentralization ideals is the use of oracles. Oracles are external data providers that supply blockchain-based smart contracts with real-world data. For instance, DeFi applications might rely on oracles to fetch the latest price of cryptocurrencies or commodities. However, the data provided by oracles is only as trustworthy as the entities that operate them. This reliance on external sources can introduce a centralized point of failure if these entities fail to provide accurate or reliable data, thereby undermining the overall decentralization of the DeFi ecosystem.

When examining Decentralized Autonomous Organizations (DAOs), a common mechanism for decision-making and governance in DeFi, we find that they often employ on-chain governance to determine the course of action. However, the decentralization of DAOs can be compromised by the fact that the funds managed by these organizations are often held in multisignature (multisig) wallets. Multisig wallets involve a select group of individuals who collectively hold the keys required to access and manage the funds. This concentration of control within a few actors can diminish the significance of on-chain governance decisions, as these few actors wield substantial influence over the DAO's assets and operations.

In summary, while DeFi is applauded for its commitment to decentralization, it is crucial to acknowledge that this decentralization varies significantly across protocols and applications. In some cases, the inherent nature of certain assets or the reliance on trusted entities in the form of custodians and oracles can introduce centralized elements into the DeFi ecosystem. Similarly, DAOs, despite their on-chain governance mechanisms, can face issues of centralization when it comes to fund management. Therefore, it becomes apparent that the term "DeFi" may not always accurately capture the intricacies of decentralization within this evolving financial landscape.

1.3 Key DeFi Risks

1.3.1 Discussion of Security Risks and Vulnerabilities Inherent in DeFi

A comprehensive exploration of the security risks and vulnerabilities inherent in the realm of Decentralized Finance is vital, as it is an emerging sector of the cryptocurrency and blockchain industry that presents both great promise and notable challenges. DeFi, at its core, represents a paradigm shift in traditional financial systems by enabling peer-to-peer transactions, lending, borrowing, and trading without the need for intermediaries such as banks or financial institutions. However, this innovative ecosystem is not without its intricacies, and understanding the potential pitfalls is crucial for participants and observers alike.

Smart Contract Vulnerabilities: DeFi platforms heavily rely on smart contracts, which are self-executing pieces of code designed to automatically facilitate and enforce agreements. However, the intricate nature of these smart contracts makes them susceptible to vulnerabilities that, if exploited, can lead to significant financial losses.

Oracles and Data Manipulation: Many DeFi applications rely on external data, oracles, to make decisions. Manipulating oracles can distort prices and trigger undesirable outcomes. This vulnerability has led to flash loan attacks, where an attacker borrows a large amount of assets, manipulates an oracle, and profits from the arbitrage opportunities created.

Centralization of Governance: Some DeFi protocols maintain a level of centralization in their governance structures, where a small group of individuals or entities have significant control. This can lead to decisions that benefit a select few but harm the broader community, potentially compromising the integrity of the platform.

Regulatory Risks: DeFi operates within a legal gray area across many jurisdictions, exposing participants to regulatory uncertainties. The looming threat includes the potential for retroactive enforcement actions, where activities once considered compliant could face legal repercussions due to evolving regulations. Regulatory fines further compound concerns, as participants may be penalized for noncompliance with ambiguous rules, potentially jeopardizing financial stability. The gravest consequence is the possibility of regulatory authorities ordering the abrupt shutdown of DeFi platforms and projects. In this ever-evolving landscape, achieving regulatory clarity remains a critical concern, demanding a delicate balance between innovation and consumer protection while necessitating active engagement with regulatory bodies to establish clearer guidelines.

User Error: DeFi's self-custodial nature empowers users with full control over private keys and security practices, but it also places immense responsibility on their shoulders. Mishandling private keys, such as misplacement or insecure storage, can result in irrevocable asset loss, similar to losing the only key to a treasure vault. Moreover, the ever-present threat of phishing attacks adds another layer of risk. Cybercriminals employ sophisticated tactics to deceive users, potentially leading to asset theft. This underscores the need for unwavering vigilance and robust security measures. In the decentralized DeFi realm, users are the sole guardians of their assets, making proper key management and cybersecurity awareness crucial for safeguarding wealth.

Interconnectedness: The DeFi ecosystem is highly interconnected, with various protocols relying on each other. If one protocol experiences a vulnerability or hack, it can have a cascading effect, impacting other connected platforms.

Economic Risks: Yield farming and liquidity provision, integral components of the DeFi ecosystem, offer enticing opportunities for users to generate returns on their crypto assets. However, these activities also introduce a layer of economic risk, especially for those who may not fully grasp the intricacies involved.

One significant source of risk is market volatility. Cryptocurrency markets are notoriously volatile, with prices subject to rapid and unpredictable fluctuations. Yield farmers and liquidity providers can find themselves exposed to significant losses if asset values plummet unexpectedly, affecting their overall returns and investments.

Moreover, DeFi protocols often employ intricate algorithmic strategies to optimize returns. Users who engage in these activities may not have a comprehensive understanding of the underlying algorithms, making them vulnerable to potential errors or vulnerabilities in the code. Such algorithmic errors can lead to substantial losses, eroding the gains users had hoped to achieve.

In navigating the DeFi space, users should exercise caution, conduct thorough research, and consider their risk tolerance carefully. While these activities offer opportunities for financial growth, they also come with the inherent risk of economic losses due to market volatility or algorithmic complexities.

Scalability Challenges: Scalability challenges within blockchain networks can have profound implications for the DeFi ecosystem, extending beyond the technical realm into the realm of accessibility and affordability. As DeFi applications gain popularity, blockchain networks can become congested, resulting in slower transaction processing times and significantly higher gas fees.

This congestion and elevated gas fee environment can make DeFi transactions prohibitively expensive, particularly for smaller investors and participants with limited resources. The cost of executing smart contracts, engaging in yield farming, or providing liquidity on decentralized exchanges can quickly add up, eroding potential gains and reducing the accessibility of DeFi to a broader user base.

The scalability issue highlights the need for ongoing technological advancements, such as layer 2 solutions and network upgrades, to address these challenges. As the DeFi space strives for greater inclusivity, scalability solutions play an important role in ensuring that the benefits of decentralized finance are accessible to a wider audience, regardless of their investment size.

In conclusion, DeFi offers exciting opportunities for financial innovation but is accompanied by a spectrum of security and operational risks. Mitigating these risks requires a combination of robust smart contract auditing, improved governance models, enhanced security practices, regulatory engagement, and user education. As DeFi continues to evolve, addressing these vulnerabilities will be crucial for its long-term sustainability and mainstream adoption.

1.3.2   The Impact of DeFi Security Breaches on Users and the Ecosystem

Security breaches in the DeFi ecosystem have significant and multifaceted impacts on both users and the broader financial landscape. These breaches introduce substantial risks that can result in immediate financial losses and long-term consequences for the ecosystem's stability and reputation.

One of the most immediate and tangible effects of DeFi security breaches is the financial losses incurred by users. When hackers exploit vulnerabilities in DeFi protocols or gain unauthorized access to user accounts, they can siphon off digital assets, leading to substantial monetary losses for affected individuals. These losses can be especially devastating for users who have invested significant sums in these platforms.

These security breaches consume the trust users place in the DeFi ecosystem. When users perceive DeFi platforms as insecure or unreliable, they may withdraw their funds and cease participating in DeFi activities. This flight of capital reduces liquidity in DeFi protocols and hinders the growth and vibrancy of the ecosystem as a whole.

The reputational damage inflicted on DeFi projects that suffer security breaches is another significant consequence. Users and the wider crypto community tend to view such projects with suspicion and skepticism. This damaged reputation can make it exceedingly difficult for these projects to rebuild trust and attract new users or investors, potentially leading to their eventual demise.

Security breaches can draw the attention of regulatory authorities, triggering a more stringent regulatory environment in the DeFi space. Governments and financial regulators may view these incidents as evidence of the need for increased oversight and regulation, impacting the freedom and flexibility that DeFi platforms initially enjoyed.

The interconnected nature of the DeFi ecosystem can magnify the consequences of security breaches. Breaches in one DeFi protocol can sometimes trigger a domino effect within the ecosystem. If an attacker successfully acquires a substantial amount of assets, they may employ these resources to attack other protocols or markets, creating systemic risks that ripple through the DeFi space.

In conclusion, DeFi security breaches have far-reaching and severe implications. Users face immediate financial losses, trust in the ecosystem is weakened, reputational damage occurs, and regulatory scrutiny may increase. Additionally, the interconnected nature of the DeFi ecosystem can amplify the impact of security breaches, underscoring the importance of robust security practices and risk mitigation measures within the DeFi space.

1.4 Overview of Top DeFi Hacks

1.4.1 Ronin

The Ronin Network breach operated in utter stealth, remaining undetected for a full 6 days following its execution. Only when a concerned user reached out to the project team, unable to withdraw approximately 5000 ETH from the project's bridge, did the alarming truth come to light. Subsequent investigations unveiled the largest DeFi breach recorded up to that point.

The Ronin Network incursion found its vulnerability through compromised private keys. Within the Ronin Network's infrastructure, a group of nine validator nodes holds the authority to validate bridge transactions, with a requisite majority of five nodes for any deposit or withdrawal. The assailant seized control of four validators held by Sky Mavis and a third-party Axie DAO validator, subsequently leveraging them to endorse their malicious transactions.

In November 2021, Axie DAO had temporarily permitted Sky Mavis to sign transactions on its behalf as a measure to assist Sky Mavis in handling an overwhelming surge of free transactions. Although this arrangement expired the following month, the allowance list was inadvertently left untouched, effectively enabling Sky Mavis to continue generating signatures for Axie DAO.

The hacker infiltrated Sky Mavis's systems and craftily exploited this allowance list to generate a signature from the Axie DAO-controlled third-party validator. Additionally, they leveraged Sky Mavis's gas-free RPC node to obtain the crucial fifth signature.

Having successfully penetrated Sky Mavis's systems, the attacker had the power to generate legitimate signatures for five Ronin Network validators. Armed with this newfound authority, they started two withdrawals, siphoning off a staggering 173,600 ETH and 25.5 million USDC from the Ronin bridge contract (Halborn, 2022b).

1.4.2   Nomad

The incident came to light through a sequence of transactions taking place on the Nomad Bridge, which connects the Moonbeam and Ethereum networks. Specifically, transactions involving the transfer of 0.01 WBTC from Moonbeam to the bridge triggered the release of 100 WBTC on the Ethereum network.

In theory, processing these transactions should have entailed a two-step procedure: first, validating the transaction's legitimacy, followed by its execution. However, the transactions directed to the bridge merely invoked the process() function within Replica.sol without prior validation. While it's possible to divide the proof and processing across multiple transactions and blocks, no preceding proof was available for these transactions.

Within the process() function, there exists an assert statement (line 185) responsible for verifying that the message associated with the transfer corresponds to a valid root. By default, an unproven message would have a root value of 0x00.

In an update to the protocol, Nomad opted to initialize the trusted root value to 0x00, and it inadvertently matched the value assigned to an untrusted root. Consequently, all messages were automatically considered as having been proven.

Once this vulnerability was uncovered, it was mercilessly exploited through a series of transactions. Even if a user did not fully grasp the underlying mechanics of why this exploit worked, taking advantage of it simply required submitting a successful exploit transaction with their own account address.

As a result of these exploits, an estimated sum of approximately $190 million was drained from the bridge (Halborn, 2022c).

1.4.3 Wintermute

The Profanity tool serves as a vanity wallet address generator, designed to create customized cryptocurrency addresses that feature easily memorable and recognizable character strings. For instance, an individual might craft a vanity address incorporating their initials or a specific word. Vanity wallet address generators like Profanity facilitate the generation of these personalized crypto addresses.

The Wintermute attack, in a departure from typical smart contract exploits, exploited a flaw in Profanity's algorithm to directly target the compromised private keys of Wintermute users.

In the realm of secure cryptographic practices, cryptographic pseudorandom number generators (CPRNGs) are typically seeded with random values to generate secure elements like private keys. However, Profanity deviated from this norm by seeding its CPRNG with a 32-bit number. Consequently, an attacker armed with substantial computing resources could employ a brute-force approach to exhaustively test possible seed values and reconstitute the private keys.

Upon the public revelation of the Profanity Hack, Wintermute did take measures to empty the ether from their hot wallet. Nevertheless, they failed to eliminate the hot wallet's address as an admin from their vault. It is probable that the private key of the hot wallet was compromised and exploited to deplete the vault. The theft amounted to $118.4 million, primarily comprising stablecoins, in addition to 671 WBTC (approximately $13 million), 6928 ETH ($9.4 million), and various other tokens. (Halborn, 2022d).

1.4.4 Poly Network

The Poly Network hack exploited a vulnerability within the Poly Network smart contracts. It is believed that the attacker managed to craft a malicious parameter, which included a fabricated block header and a forged validator signature. This manipulation allowed the attacker to circumvent the bridge's standard validation process and execute token withdrawals from the bridge directly to their own address.

This attack encompassed 57 different cryptocurrencies spanning across 10 distinct blockchains. The attacker was able to make off with an estimated $610 million from the protocol. (Halborn, 2023).

1.4.5 Euler Finance

The hacker orchestrated a complex series of maneuvers involving three contracts: a primary contract, along with two others designated for violation and liquidation purposes. Here's a breakdown of their actions:

The hacker initiated a flashloan of 30 million DAI from Aave, a flashloan protocol, and directed these funds to the violation contract.

Subsequently, they deposited 20 million DAI into Euler Protocol and received approximately 19.6 million eDAI in return.

With the acquired 19.6 million eDAI, the hacker borrowed approximately 195.6 million eDAI and 200 million dDAI.

It's worth noting that the hacker still had ten million DAI left from the initially borrowed 30 million DAI. They used this remaining ten million DAI to partially repay their debt. This step was crucial because the Euler Finance smart contract assessed the health score of borrowing accounts. The balance at this stage was 190 million dDAI. Following this, they borrowed an additional 195.6 million eDAI and 200 million dDAI.

At this point, the attacker executed a donation of 100 million eDAI to the Euler protocol reserve. This operation succeeded because the donateToReserve function lacked a liquidity check.

The subsequent liquidation call proved successful for the attacker, resulting in them receiving 254 million dDAI and 310 million eDAI. They promptly repaid to Aave the 30 million DAI borrowed initially and generated approximately 8.7 million DAI in profit from the exploit.

Not content with this success, the attacker employed the same modus operandi with WETH.

These intricate maneuvers allowed the hacker to exploit system vulnerabilities and amass substantial gains (Hacken, n.d.).

1.4.6   Qubit Finance

The attacker initiated their scheme by invoking the deposit function in the QBridge ETH contract, using malicious data that managed to pass all of the contract's validation checks. Notably, the transaction they submitted contained no ETH.

Within this function, the contract called the safeTransferFrom function within the token's contract to facilitate the transfer of tokens from the depositor into the contract. This call to safeTransferFrom would typically result in the deposit function failing if the provided token address was invalid.

However, the critical flaw in this scenario was that the contract did not utilize OpenZeppelin's SafeERC20 library. Had the contract employed this library, the exploit would have been thwarted because SafeERC20.safeTransferFrom incorporates the functionCall() (a function from OpenZeppelin's Address.sol contract), which verifies the existence of contract code at the target address. In the case of the 0 address, this check would fail as it has no associated code.

However, the contract in question utilized a modified safeTransferFrom() function that directly employed the call() function. Since the 0 address lacks any code, no code execution occurred, and the call succeeded without reverting. Consequently, the deposit function executed without issues, but no actual tokens were deposited.

The Ethereum QBridge protocol, misinterpreting this as a legitimate Deposit event, considered it a valid deposit of ETH. Consequently, qXETH tokens were minted for the attacker on the Binance Smart Chain (BSC).

By repeating this process multiple times, the attacker managed to accumulate a substantial quantity of qXETH tokens without depositing any real tokens into the protocol. Subsequently, they successfully converted these tokens into BNB, ultimately draining approximately $80 million in assets from the protocol. (Halborn, 2022a).

1.4.7   Curve

Curve Finance operates as a decentralized finance protocol that enables users to swap stablecoins on the Ethereum blockchain without the need for intermediaries. The protocol relies on liquidity pools, where participants pool their assets within smart contracts.

During the hack, certain stablecoin liquidity pools within Curve Finance became targets due to vulnerabilities present in the coding language utilized by these pools. Specifically, the affected pools were built using Vyper, a third-party programming language designed for Ethereum smart contracts. Although Vyper had undergone upgrades in the past, some of the older versions, notably version 0.2.15, were still in use by the liquidity pools targeted in the attack.

In an initial tweet from Curve Finance, the team attributed the hack to a malfunctioning reentrancy lock. However, they also noted that a comprehensive investigation was ongoing to gain a precise understanding of the events leading up to and during the breach.

Among the stable pools impacted were Metronome's msETH/ETH pool, which at the time had been drained of approximately $3.4 million; the Curve DAO, which saw losses of around $24.7 million; PEGD's pETH/ETH pool, which suffered a loss of $11 million; and Alchemix's alETH/ETH pool, which was drained of $22.6 million at the time of reporting.

Additionally, there were reports of similar attacks occurring on the Binance Smart Chain (BNB), resulting in a loss of up to $78,000. These attacks targeted pools that utilized the Vyper programming language, while the remaining pools on Curve Finance remained unaffected and, for now, secure (TrustWallet, n.d.).

Sincere gratitude is extended to c0ffeebabe.eth, an MEV bot that intercepted a hacker's transaction and subsequently returned the tokens to Curve and its users. The security of web3 is enhanced by the efforts of entities like c0ffee-babe.eth.

1.4.8 Summary of the Hacks and Causes

Finding a single root cause for DeFi hacks is an elusive task, but identifying the most likely culprits is feasible. These primary contributors include:

Insecure Private Key Management: The leading cause revolves around the failure to safeguard private keys. This vulnerability has precipitated numerous bridge exploits and poses substantial risks, echoing the tactics of web2 hacks employing social engineering.

Composability Issues Between Smart Contracts: The secondary cause arises when two separate smart contracts operate smoothly independently but trigger complications when they interact. An illustrative yet not entirely precise example is the flash loan attack. While the flash loan contract and the end contract may function adequately in isolation, the accessibility of substantial funds through flash loans renders the end contract susceptible, even though it might not have been secure to begin with.

Logic Flaws in Smart Contracts: The third common cause originates from errors in the logic of a contract. Such instances, reminiscent of the Ethereum DAO hack, have become less frequent over time as auditors improve their practices. Nevertheless, they remain potent threats.

However, the DeFi landscape is populated with countless other reasons for hacks and failures. Consider the case of Curve Finance in July 2023, where the smart contract wasn't inherently flawed; instead, it was the compiler used to compile the contract code that was compromised. Similarly, Mango Market operated smoothly until October 2022 when the mango token's price was manipulated, leading to financial instability in the money market.

The DeFi realm faces a plethora of existing challenges, with many more awaiting discovery. For instance, a novel reentrancy attack known as the "read-only reentrancy attack" emerged in 2022. New attack vectors will likely emerge, either because they are presently impossible but will become feasible or because they currently elude auditors and hackers. This dynamic landscape underscores the need for ongoing vigilance and innovation in the DeFi sector.

1.5 Defense Measures

1.5.1 Importance of Security Audits

Security audits are fundamental in the world of smart contracts, playing a crucial role in safeguarding digital assets and fortifying the resilience of decentralized applications (DApps) and blockchain ecosystems. These audits are a miscellaneous approach that dives deep into the complexity of the code, transcending mere technicalities to impact various aspects of the blockchain landscape profoundly.

At the heart of smart contracts lies their codebase, the very foundation upon which these contracts operate autonomously. Without human intervention, they execute complex transactions, making code integrity an absolute necessity. Any compromise in code integrity opens the door to vulnerabilities that malicious actors can exploit, putting assets and platform trust at grave risk. Security audits, in their meticulous scrutiny, identify and rectify these vulnerabilities, standing as guardians of the contract's integrity.

Smart contracts oversee significant financial transactions and digital assets, making them prime targets for cyberattacks. Here, security audits act as proactive defenders, mitigating the looming risk of financial losses due to security breaches. By addressing vulnerabilities preemptively, these audits construct a fortified environment for users, assuring their assets remain safe and sound.

In the blockchain realm, trust derives from the code itself, rather than traditional legal agreements. A security breach not only endangers assets but also corrodes the very trust and credibility upon which the contract and the underlying blockchain network are built. Security audits are the bedrock of confidence, assuring users that the contract has undergone rigorous testing and is resilient against adversarial actions.

Vulnerabilities unearthed in security audits are essentially potential entry points for attackers. Addressing these issues before deployment, security audits form an essential barrier against exploitation, shoring up the contract's defenses and reducing susceptibility to hacking or manipulation.

In the world of DApps and blockchain platforms, user participation is crucial. Users seek assurance that their assets and transactions are secure in this decentralized landscape. Security audits, with their seal of approval, provide this much-needed reassurance, drawing in a wider user base and propelling the adoption of blockchain-based applications to new heights.

Security audits are not just a checkpoint but the backbone of the smart contract ecosystem. They ensure the integrity of code, mitigate risks, and instill user confidence. As proactive measures, they ward off exploitation, ultimately reinforcing the trustworthiness and resilience of smart contracts within the ever-evolving blockchain landscape.

1.5.2 Smart Contract Best Practices

To ensure the security of your smart contract, it's crucial to adopt best practices that encompass various aspects. Firstly, be prepared for failure by implementing mechanisms to pause or respond to issues promptly. Develop a robust upgrade strategy to address bugs and vulnerabilities and manage the financial risk effectively.

Careful rollouts are essential to detect and resolve bugs early. This involves thorough testing, incremental deployment, bug bounties, and continuous testing for new attack vectors.

Simplicity is key to reducing errors in your contracts. Keep the logic simple, reuse existing code, and modularize your functions. Utilize blockchain decentralization only where necessary and prioritize clarity over performance.

Stay updated with security developments and regularly audit your contracts. Embrace new security techniques and promptly upgrade tools and libraries to their latest versions.

Be mindful of blockchain properties, especially regarding external contract calls, data visibility, public function execution, imprecise timestamps, gas limits, and random number generation.

Consider fundamental trade-offs between software engineering and security. Balance modular, upgradeable components with code duplication and monolithic designs based on your contract's complexity and lifespan.

These practices help developers approach smart contract security with a comprehensive mindset, preparing them to navigate the specific challenges posed by Solidity and Ethereum development.

1.6 Conclusion

This chapter has provided an overview of decentralized finance and highlighted the vital role of security in this financial revolution.

DeFi has proliferated, reshaping the way people manage their finances. It differs from traditional banking and is accessible to a broader audience. However, due to its decentralized nature, it faces unique security challenges.

We've explored the risks and vulnerabilities within the DeFi space, emphasizing their potential impact on both individuals and the entire system. We've also examined real world examples of DeFi hacks to illustrate the consequences of security breaches.

On a positive note, we've discussed how security audits and best practices can enhance safety in DeFi. While they do not offer absolute certainty, these measures can reduce the likelihood of problems.

In conclusion, this chapter has highlighted the collective responsibility within the DeFi ecosystem to prioritize security and remain well-informed. This commitment contributes significantly to the continued growth and evolution of DeFi as a transformative influence within the financial landscape.

References

Halborn. (2022a, January). Explained: The Qubit Hack. Halborn. https://www.halborn.com/blog/post/explained-the-qubit-hack-january-2022

Trust Wallet. (n.d.). The Curve Finance Hack Explained. Trust Wallet Blog. https://blog.trustwallet.com/blog/the-curve-finance-hack-explained

Hacken. (n.d.). Euler Finance Hack. Hacken. https://hacken.io/discover/euler-finance-hack/

Halborn. (2022b, March). Explained: The Ronin Hack. Halborn. https://www.halborn.com/blog/post/explained-the-ronin-hack-march-2022

Halborn. (2022c, August). Explained: The Nomad Hack. Halborn. https://www.halborn.com/blog/post/explained-the-nomad-hack-august-2022

Halborn. (2022d, September). Explained: The Wintermute Hack. Halborn. https://www.halborn.com/blog/post/explained-the-wintermute-hack-september-2022

Halborn. (2023, July). Explained: The Poly Network Hack. Halborn. https://www.halborn.com/blog/post/explained-the-poly-network-hack-july-2023

Carlo Parisi is a senior Solidity smart contract developer, senior auditor, and content creator with a degree in Computer Science. He has a deep knowledge in developing and auditing Solidity code, the main language used in Ethereum for smart contracts. He has been a Bitcoin Enthusiast since 2013, DeFi user since 2018.

Dmitriy Budorin is Founder & CEO at Hacken and Founder at HackenProof.

Dyma is a cybersecurity expert and crypto economy influencer with 14+ years of managerial expertise in cybersecurity as well as risks and controls audits. Dyma holds a master's degree in International Economics and an MBA from the Kyiv Institute of Investment Management. He is a certified member of the Association of Chartered Certified Accountants (ACCA).

In 2017 Dyma established Hacken, a cybersecurity consulting firm. Five years later, Hacken is trusted by the largest crypto projects; the company's portfolio includes HackenAI, HackenProof, CER, and a suite of accompanying blockchain services. Dyma's effective leadership is what transformed Hacken from a startup into a major player in Web3 cybersecurity. The story of success is only gaining momentum.

As the company's Co-Founder and CEO, Dyma is responsible for leading the team of 100+ talented specialists and providing a vision of the future. Dyma consults the Ukrainian government on the adoption of a virtual economy. He is a regular participant in major Web3 cybersecurity events as an invited speaker.

Chapter 2
NFT Security

Lisa JY Tan

Abstract This chapter provides a comprehensive overview of the security landscape surrounding non-fungible tokens (NFTs). It examines the unique attributes of
NFTs that present both opportunities and risks, such as immutability, provenance,
and trustless transactions. Different categories of threats are analyzed, including
exchange vulnerabilities, smart contract exploits, social engineering tactics, and
issues with digital storage solutions like IPFS. Real-world case studies of major
security incidents are detailed, covering the financial, reputational, and ecosystem
impacts. Mitigation strategies spanning technological safeguards, regulatory measures, and community engagement are outlined as crucial components in securing
the burgeoning NFT ecosystem.

2.1 Introduction to NFT Security

As NFTs increasingly permeate the digital economy, their unique attributes—
immutability, provenance, and the ability to tokenize nearly any asset—present both
unprecedented opportunities and significant security challenges. In volume I of this
book series, we covered the design challenges and framework to create tokens. This
chapter introduces the reader to the foundational aspects of NFT security, covering
the role of IPFS in storage, the potential risks of smart contract vulnerabilities, and
the insidious nature of social engineering within the marketplace. It sets the stage
for a thorough exploration of the security incidents that have shaped the current
understanding of NFT security and the measures that can be taken to mitigate risks
and protect the ecosystem.

L. JY. Tan (✉)
Economics Design, Singapore, Singapore
e-mail: lisa@economicsdesign.com

© The Author(s), under exclusive license to Springer Nature
Switzerland AG 2024
K. Huang et al. (eds.), *Web3 Applications Security and New Security Landscape*,
Future of Business and Finance, https://doi.org/10.1007/978-3-031-58002-4_2

2.1.1   About NFT

Non-fungible tokens (NFTs) represent a revolutionary type of digital asset, each with unique characteristics that are verifiably owned and tracked on a blockchain, such as Ethereum. Think of it as a digital certificate of record that is unique and immutable as it exists on a blockchain.

These tokens can be connected to both tangible and intangible assets, serving as an advanced "certificate of ownership" for the asset they represent. This allows for the tokenization and tracking of physical items, making ownership backgrounds easily verifiable. NFTs can encapsulate anything imaginable, from art and sports memorabilia to memes and virtual real estate, thereby enabling a wide range of assets to be bought, sold, and traded in digital form.

The key features that make NFTs particularly attractive include their ability to confer clear ownership, their traceability, and the trustless nature of their transfer, which allows for transactions between parties without mutual trust. The technology underpinning NFTs not only encourages the further digitization of the economy but also addresses previous blockchain limitations by facilitating the creation, storage, and trading of these tokens.

The trajectory of NFTs saw a significant uptick beginning in 2014 with the creation of "Quantum" by McCoy (2014), but it wasn't until 2017 that NFTs began to capture widespread attention. This period marked the emergence of numerous NFT collections on the Ethereum blockchain, which offered solutions to the challenges faced by NFTs on earlier blockchain platforms, such as issues with trading and ownership transfers. The advent of Ethereum significantly lowered the barriers to entry for NFT creators and collectors.

2.1.1.1   Rise of NFT

The rise of NFTs was further propelled by two major factors before 2021: the COVID-19 pandemic, which drove more people toward NFT communities online, and several high-value NFT sales that captured public interest.

During the COVID-19 pandemic, Axie Infinity revolutionized the NFT space by popularizing the play-to-earn model, becoming the first Ethereum blockchain-based online video game that incorporates NFTs with unique attributes for gameplay and trading. This model not only garnered significant traction, inspiring a wave of NFT-based games, but also faced criticism for its gameplay and payout system, likened by some to gambling (Bloomberg, 2021). Here, the NFT represents digital and virtual assets, a character in the game called Axie The rise of Axie (2021).

Notably, the artist Beeple sold an NFT for $69 million through a major auction house, a sale that not only underscored the potential value of digital art but also highlighted the expanding appeal of NFTs (BBC News, 2021). Following Beeple's sale, other notable transactions included Edward Snowden's "Stay Free" (Snowden, n.d.) and a CryptoPunks NFT, each fetching millions. The explosion of interest in digital art and collectibles throughout 2021, alongside the emergence of NFT-based virtual worlds and games, has solidified NFTs as a pivotal innovation in digital ownership and the broader web3 landscape, with major brands and companies increasingly exploring NFT projects. Here, the NFT presents digital art.

NFT marketplaces such as OpenSea, Rarible, and Axie play a crucial role in the NFT ecosystem by providing the necessary infrastructure for the distribution and trading of NFTs, thereby fueling the interest of crypto art collectors and traders alike. In September 2021, the combined all-time trading volume of the top three marketplaces—OpenSea, Axie, and CryptoPunks—exceeded $10 billion USD, with individual NFT sales reaching record highs (CryptoBriefing (n.d.)). High-profile sales, such as digital artist Beeple's artwork fetching $69.3 million USD and Twitter CEO Jack Dorsey's first tweet selling for $2.9 million USD, underscore the booming interest and financial stakes in the NFT market.

By January 2022, OpenSea, the largest NFT marketplace, generated more than $5 million USD in trading volume, showcasing the immense economic activity within this space (Benson, 2021a). Despite OpenSea's high trading volume, January saw users experiencing significant security risks, with high-value NFT collectibles being sold for a fraction of their estimated worth due to a UI exploit, leading the firm to compensate affected customers with $1.8 million worth of ETH (Benson, 2021b).

2.1.2 Significance of Security in NFTs

Security in the NFT space is paramount due to the unique and irreplaceable nature of these digital assets, which often carry significant financial value and are a testament to the digital identity and reputation of their creators and owners. The NFT ecosystem's reliance on blockchain technology offers inherent security features like verifiability and trustless transactions, facilitating the safe transfer of assets between parties without mutual trust. However, the ecosystem is not immune to security vulnerabilities and scams, such as impersonation, counterfeit NFTs, and unauthorized sales, which can lead to substantial financial losses and damage to reputations (Das et al., 2022). The potential for such breaches underscores the critical need for robust security measures and vigilant practices within NFT marketplaces and among users to protect against financial and reputational harm.

In the realm of non-fungible tokens (NFTs), the security of transactions and the integrity of the assets being traded are paramount, not only to preserve the financial investments of buyers but also to maintain the credibility of the marketplace. A stark illustration of the dire need for enhanced security measures within the NFT space is the Frosties collection rug pull scam. A "rug pull" is a type of scam where developers promote a project—often with significant hype—to drive up the value of associated NFTs or cryptocurrencies, only to suddenly withdraw all the funds from the project, leaving investors with worthless assets. In the case of the Frosties scam, investors were enticed into buying into a seemingly promising NFT collection, only for the creators to disappear with approximately $1.3 million in Ethereum, the cryptocurrency used for the transaction (Geron, 2023). This incident not only resulted in substantial financial losses for the investors but also inflicted reputational damage on the broader NFT market, highlighting the critical importance of conducting due diligence and implementing robust security protocols to protect against such fraudulent schemes.

Another case specific to an individual than an entire community, Todd Kramer, an NFT collector, serves as a stark reminder of the financial and reputational stakes involved. Kramer fell victim to a phishing attack, resulting in the loss of 16 valuable NFTs, including eight "Bored Ape" NFTs, totaling a loss of approximately 593 ETH, or about $1.7 million at the time (CryptoBriefing (n.d.)). This incident not only highlights the direct financial impact of security breaches but also underscores the reputational risks for platforms involved. OpenSea's intervention to freeze the stolen assets and prevent their trade on its platform was a remedial action that, while beneficial, also sparked debate about the centralization of power in an ecosystem that values decentralization. This case vividly illustrates the multifaceted impact of security breaches in the NFT space, encompassing financial loss, reputational damage, and the broader implications for the principles underpinning the decentralized digital economy (Gupta et al., 2022).

2.1.3   Common Security Threats

In the burgeoning NFT marketplace, the security of transactions and authenticity of assets is paramount due to the high value and digital nature of these assets. A glaring example of the security threats in this space is highlighted by the incident involving a collector known as Pranksy, who was deceived into buying a fake Banksy NFT for £244,000 (Hern, 2021). This incident underscores the vulnerabilities within the NFT market, particularly around the authenticity and verification of digital assets, and the ease with which fraudsters can exploit the excitement around NFTs to conduct sophisticated scams, leading to significant financial losses for unsuspecting investors.

Another critical security threat is identified through the exploitation of technical vulnerabilities within NFT marketplaces, as detailed in Gupta et al. (2022) paper. A notable incident involved OpenSea, where a vulnerability allowed attackers to potentially steal a user's entire crypto wallet by sending a malicious NFT. This exploit, taking advantage of OpenSea's storage domain vulnerabilities, could lead to the loss of all funds in the victim's wallet (OpenSea NFT marketplace hacked (n.d.)). Such vulnerabilities highlight the essential need for robust security practices and the risks associated with human interaction with technology, emphasizing the importance of Layer-8 security measures that account for human factors in the security architecture (Kale et al., 2021).

Further examination of security challenges in NFT marketplaces reveals various security, privacy, and usability issues, including those related to user authentication and the transparency of marketplace contracts (Das et al., 2022). The study underscores the complexity of ensuring secure transactions within these platforms, where weaknesses in design or implementation can result in substantial financial losses. The analysis of incidents and vulnerabilities within marketplaces such as OpenSea, Rarible, and others demonstrates the multifaceted nature of security threats, from technical exploits to social engineering tactics, and underscores the critical need for ongoing vigilance, comprehensive security strategies, and user education to safeguard the NFT ecosystem from malicious actors.

2.1.4   Notable Breaches

In the dynamic realm of non-fungible tokens (NFTs), the security incidents witnessed have underscored the nascent technology's growing pains and the urgent need for reinforced cybersecurity measures. One of the most significant security breaches in the NFT space is the Cross-Chain DeFi Site Poly Network Hack, which sent shockwaves through the crypto community. In this event, attackers exploited vulnerabilities in the smart contract code of the Poly Network platform, orchestrating a heist of approximately $610 million in cryptocurrency (Reuters, 2021).

The hack was made possible by a mismanagement of access rights between two important Poly smart contracts, EthCrossChainManager and EthCrossChainData. The hacker was able to take advantage of this vulnerability to transfer the funds to external wallet addresses. Following the attack, Poly Network urged the hacker to return the funds, and the hacker started returning the tokens over the following days. The hacker claimed that the purpose of the theft was to reveal vulnerabilities and secure Poly Network. While almost all of the stolen assets have been returned, a portion of the funds is still locked in an account that requires passwords from Poly Network and the hacker.

This event not only marked one of the largest thefts in DeFi history but also cast a spotlight on the critical vulnerabilities in smart contract security, demonstrating the sophisticated methods used by cybercriminals to exploit weaknesses in blockchain technology The theft of NFTs (2022).

Another significant incident involved OpenSea. Users were exposed to an insidious vulnerability (Dent, 2021) that could have resulted in the complete loss of their crypto wallets due to a malicious NFT. This vulnerability, which originated from a deceptive storage domain, could have allowed attackers to access and drain a victim's entire wallet. OpenSea's quick response to fix the vulnerability and their pledge to increase security education efforts demonstrate the critical need for platforms to adopt proactive security measures. These incidents pave the way for a deeper analysis of the inherent risks within the NFT ecosystem and underscore the necessity for robust security frameworks and rigorous threat assessments to safeguard against future attacks.

2.1.5 The Need for Enhanced Security Measures

The security of digital assets in the NFT space is of paramount importance, as the integrity of transactions and the authenticity of assets directly impact both the financial value and the trust within the ecosystem. Security breaches, such as theft, piracy, and copyright infringement, degrade the uniqueness and value of NFTs, leading to significant financial and reputational damage. For example, over $100 million worth of NFTs were reported as stolen through various scams from July 2021 to July 2022, with July 2022 marking a record high of over 4600 stolen NFTs, indicating that security threats persist even during the crypto bear market. This highlights the urgent need for robust and scalable security measures to protect NFT assets and platforms from such illicit activities (Elliptic NFT Report, 2022).

The case of the Poly Network hack serves as a case study underscoring the vulnerabilities in smart contract security, where attackers exploited code vulnerabilities to steal about $610 million in cryptocurrency. This incident, among the largest in DeFi history, signifies the potential scale of impact a security breach can have, necessitating the implementation of rigorous security measures to safeguard assets and maintain market integrity. In response to these threats, solutions like Intertrust's MarketMaker, which integrates ExpressPlay's multi-DRM solution, aim to enhance NFT security beyond the blockchain (Intertrust). This addresses critical content protection and copyright issues, aiming to preserve the long-term value of NFTs against unauthorized access and distribution, thus maintaining their exclusivity and worth in the marketplace (Introducing MarketMaker (n.d.).

2.2 Exchange Vulnerabilities and Exploits

2.2.1 Mechanisms of Exchange Exploits

Attacks on NFT exchanges exploit both technical vulnerabilities and the human element, leading to unauthorized transactions and price manipulations that can severely impact the financial structure and reputation of the platforms involved. A poignant case study exemplifying this is a credential-stuffing attack on OpenSea. Attackers gained unauthorized access to user accounts, resulting in the theft of NFTs and cryptocurrency tokens, worth $1.7 million (Paganini, 2021). The forensic investigation revealed extensive monetary losses and the movement of assets between wallets, painting a clear picture of the aftermath of such breaches.

From a technical standpoint, a user experienced a security issue where attackers could potentially steal their entire crypto wallet by sending a malicious NFT. This exploit, relying on fraudulent pop-up URLs from the platform's storage domain, could have allowed attackers to drain a victim's funds entirely, underscoring the pressing need for robust security frameworks to prevent exploitation. These incidents reveal that security vulnerabilities can be exploited to perform illicit activities such as unauthorized asset transfers and price manipulation, which not only result in financial loss for users but also erode trust in the marketplace, emphasizing the necessity for vigilant security measures and user education to mitigate these "Layer-8 risks." Layer-8 risks mean users in the cybersecurity world (Kale et al., 2021).

2.2.2 Exploiting Vulnerabilities in NFT Marketplaces

Attackers exploit vulnerabilities in NFT exchanges using sophisticated methods that compromise the integrity of transactions. By targeting weak points in the NFT Marketplaces such as insufficient security protocols or system flaws, attackers can manipulate NFT prices and execute unauthorized transactions. These nefarious activities can include injecting malicious code through NFTs, as seen in the OpenSea incident discussed above, where pop-ups from a fraudulent URL could allow attackers to access and siphon funds from a victim's crypto wallet. Such vulnerabilities open the door not only to financial theft but also to manipulating the perceived value of NFTs, destabilizing the marketplace, and eroding user confidence.

In a disturbing case study, a major NFT exchange suffered a credential-stuffing attack, where attackers used stolen login credentials to gain unauthorized access to user accounts. This type of attack leverages large databases of compromised user information, often obtained from breaches of other services. The consequences of this security breach were severe, with the theft of valuable NFTs and cryptocurrency tokens from multiple accounts. The subsequent forensic investigation revealed the extent of the damage, including a significant monetary loss for the victims and the illicit movement of assets, underscoring the critical need for exchanges to bolster their authentication processes and monitor for unusual account activity to prevent such devastating exploits. The details of this incident highlight the urgent necessity for continual enhancement of security measures to protect against the ever-evolving threat landscape in the NFT domain.

Another notable incident was the NFT Trader platform,[1] a service facilitating the exchange of NFT assets, fell victim to a critical exploit. This security breach was notable for its use of a smart contract vulnerability that allowed an attacker to take possession of valuable NFTs without proper authorization (NFTNow, 2021). The exploit involved outdated smart contracts that hadn't removed previous authorizations, thereby enabling the unauthorized acquisition of assets.

The resolution of this exploit was as remarkable as the attack itself, involving the collaborative efforts of the NFT community. Most notably, a 16-year-old developer, who goes by the name "0xNuclear," played a pivotal role in identifying and mitigating the exploit. The young developer, with the aid of the platform's community, was able to analyze the faulty code and propose a solution to halt the ongoing theft,

---

[1] NFTTrader.io.

showcasing the strength of community vigilance and expertise in the face of security challenges.

The potential damages from this exploit were substantial, with the possibility of
millions of dollars worth of NFTs at risk. While the exact financial impact is challenging to ascertain due to the varying valuations of NFTs, the incident served as a
stark reminder of the potential vulnerabilities within smart contracts and the need
for constant monitoring and rapid response mechanisms. It also highlighted the
importance of community-driven security and the value of skilled individuals,
regardless of age, in safeguarding the NFT ecosystem.

2.2.3   NFT Platform Insider Trading

The trust users place in NFT marketplaces is paramount to the stability and growth
of the digital asset market. However, vulnerabilities and unethical practices within
these platforms can severely erode this trust. A case in point is the insider trading
scandal at OpenSea, yet again do we reference OpenSea in our case studies.

Nathaniel Chastain, a former product manager at OpenSea, exploited his access
to confidential information about which NFTs were to be featured on the platform
for personal gain (Yirka, 2022). By purchasing these NFTs before they were publicly featured, knowing that their value would likely increase once highlighted, and
then selling them at a profit, Chastain engaged in a form of insider trading that is
illegal in traditional financial markets.

This betrayal of trust resulted in charges of wire fraud and money laundering
against Chastain, as his actions not only violated legal statutes but also the ethical
standards expected within the marketplace. The case brought to light the susceptibility of the NFT market to traditional financial crimes, and the subsequent legal
actions underscored the seriousness with which such transgressions are treated. The
repercussions of this incident were not limited to the legal consequences faced by
the perpetrator; they also cast a shadow over the perceived security and trustworthiness of NFT platforms. Users' confidence was shaken, and the market's stability
was challenged, as it became evident that the mechanisms in place were insufficient
to prevent such abuses of power and access. This incident highlighted the urgent
need for robust and transparent security measures to prevent insider trading and
protect the integrity of the NFT marketplace.

2.2.4   Security Measures for Exchanges

To bolster security and prevent incidents similar to those that have shaken user trust
in NFT marketplaces, it is essential to establish and adhere to rigorous security
protocols and practices. These protocols must be comprehensive, encompassing
both technical measures and operational policies. From a technical standpoint,

deploying smart contracts that have undergone extensive audits by reputable third-party firms can identify vulnerabilities before they are exploited. Additionally, incorporating real-time monitoring systems to detect unusual trading patterns or access anomalies can help in quickly identifying and mitigating unauthorized activities. Implementing robust authentication processes, including multi-factor authentication and routine security training for employees, can further reinforce the defense against insider threats. We will discuss smart contract flaws and exploit in the next segment.

On the operational front, establishing clear and enforceable policies around confidentiality and data access within an organization is critical. Employees should be aware of the ethical and legal implications of misusing insider information, and there should be a transparent mechanism for reporting suspicious activities. Moreover, the practice of "least privilege" should be enforced, where employees only have access to the information necessary to perform their duties. Regular security briefings and updates can keep the team informed about the latest threats and security best practices. These practices, along with the continuous evolution of security measures to address new and emerging threats as suggested by the research community at Smart Contract Research Forum, create a robust framework to safeguard against future security breaches in the NFT ecosystem (Das et al., 2022).

2.3 Smart Contract Flaws and Exploitations

Smart contracts are the bedrock of NFT transactions, encoding the rules and execution of exchanges within the blockchain. However, their strength is also their Achilles' heel; they are prone to security breaches if not meticulously written and tested. Vulnerabilities in smart contracts often arise from common coding flaws that can be exploited by bad actors. Reentrancy attacks, where a function is externally called before the first execution is completed, can drain funds from a contract. Integer overflow and underflow can result in incorrect calculations, leading to unanticipated outcomes. Improper access control can inadvertently open up administrative functions to unauthorized users. These vulnerabilities are not just coding errors but fundamental oversights that can compromise the integrity of the blockchain and result in significant financial losses.

One of the most infamous smart contract breaches is The DAO Hack, a textbook case of a reentrancy attack that led to the theft of 3.6 million Ether, shaking the Ethereum community to its core (Blockgeeks, n.d.). The decentralized autonomous organization (DAO) was built to act as a venture capital fund for the crypto and decentralization space, but due to a reentrancy vulnerability in its smart contract, attackers were able to recursively withdraw Ether. This hack not only resulted in substantial financial loss but also raised questions about the security and viability of smart contracts, leading to a hard fork in the Ethereum blockchain. The DAO Hack remains a cautionary tale that underscores the need for rigorous smart contract auditing and security measures to prevent similar vulnerabilities from compromising smart contracts in the future.

2.3.1 Examples of Smart Contract Exploits

The digital asset realm, specifically the Non-Fungible Token (NFT) and DeFi sectors, has experienced notable security breaches due to vulnerabilities in smart contracts. These incidents highlight the urgent need for robust security measures and comprehensive audits to safeguard digital properties. Three significant real-world cases exemplify the impact of such vulnerabilities:

1. NFT Trader Contract Compromise: The NFT trading platform, NFT Trader, fell victim to a security breach involving two of its older contracts. This compromise led to the theft of a range of valuable NFTs. The incident underscores the necessity for continuous monitoring and updating of smart contracts to prevent vulnerabilities from being exploited (Crypto News, 2024).
2. Wyvern Protocol Weakness: The Wyvern Protocol, a foundational element in NFT exchanges, was exploited by hackers. They manipulated the protocol to gain unauthorized access to NFTs owned by various users. The exploit involved eliciting user authorization in partially completed smart contracts and subsequently filling in the remaining sections to facilitate unauthorized transfers. This case illustrates the risks associated with smart contract interfaces and the importance of ensuring secure user interaction mechanisms (4irelabs, 2023).
3. Beanstalk DeFi Exploit: The decentralized finance (DeFi) protocol Beanstalk suffered a significant exploit, showcasing the potential misuse of flash loans to manipulate the market and exploit smart contract weaknesses.

The attack exploited Beanstalk's governance protocol, allowing the attacker to extract funds into a private Ethereum wallet. The attacker used a flash loan obtained through the decentralized protocol Aave to borrow a significant amount, which provided them with the voting power to approve the execution of code that transferred the assets to their own wallet. The entire process, including the loan and the attack, took place in less than 13 s. The attack was made possible by the implementation of the Curve LP Silos, which ultimately permitted the attacker to conduct an emergency execution of a malicious proposal. The stolen assets were then liquidated into Ethereum, and approximately $76 million in non-Beanstalk assets were stolen from liquidity pools. Flash loans are uncollateralized loans that allow users to borrow assets without providing upfront collateral. They are typically used within the decentralized finance (DeFi) ecosystem to perform various financial activities within a single transaction block, such as arbitrage maneuvers.

This incident highlights the broader implications of smart contract vulnerabilities, extending beyond NFTs to impact the DeFi sector at large (Hacken, 2023).

These cases collectively underscore the critical importance of implementing stringent security protocols and conducting thorough audits within the blockchain and NFT domains. They serve as a stark reminder of the ongoing battle against vulnerabilities in smart contracts and the need for continuous vigilance and improvement in security practices.

2.3.2   Mitigation Strategies

To mitigate the risks associated with smart contract development and deployment, several strategic measures are recommended. The implementation of coding standards and security patterns is crucial. These provide a structured approach to coding, helping developers avoid common pitfalls and vulnerabilities inherent in smart contract development. Additionally, leveraging tools like OpenZeppelin, which offers secure, standard smart contract libraries, aids developers in constructing robust contracts. These libraries have been thoroughly tested and audited, significantly reducing the chance of vulnerabilities (OpenZeppelin Security Audits (n.d.)). Furthermore, actively engaging the wider community through reviews and bounty programs is vital. These programs incentivize the discovery and reporting of vulnerabilities, turning potential security weaknesses into opportunities for improvement and community engagement.

In practice, OpenZeppelin's role in the ecosystem serves as an exemplar of these mitigation strategies. By providing a suite of audited, community-vetted smart contract libraries and implementations, OpenZeppelin reduces the barrier to secure contract deployment. Projects utilizing OpenZeppelin can inherit from well-established contract templates, ensuring that they are built upon a foundation of code that conforms to the latest security standards. OpenZeppelin's security audits, conducted by experts in the field, further contribute to this ecosystem, offering an additional layer of scrutiny that bolsters the security posture of numerous projects. These methods collectively form a multifaceted approach to risk mitigation, addressing smart contract security from both the technical and community angles, which is essential for fostering a secure blockchain environment.

2.4   Social Engineering Tactics in the NFT Space

Social engineering is a manipulation technique that targets the human element of security systems. It is particularly effective in the NFT market because it exploits human psychology to gain confidential information, unauthorized access, or assets fraudulently. The principles of social engineering hinge on the attacker's understanding of human behavior, leveraging tactics such as impersonation and fraud. These tactics are especially potent in the NFT space, where enthusiasm and the Fear Of Missing Out (FOMO) can lead investors to make hasty decisions without proper diligence, making them susceptible to fraud. NFT platforms, such as OpenSea, have experienced issues where attackers could potentially steal a user's entire crypto wallet by sending a malicious NFT, a tactic that relies on deceiving users with fraudulent pop-ups from the platform's storage domain (Cryptobriefing).

Applying these principles to the NFT market, we see impersonation and fraud executed through sophisticated phishing schemes, where attackers automate low-effort tasks to target a broad range of users, including high-profile NFT owners. A notable case involved an NFT collector who fell victim to a phishing contract, resulting in the theft of several valuable NFTs (NFT collector loses (2022)). The attack's success and subsequent asset freezing by OpenSea raise questions about the centralized power within NFT platforms and the need for clear ownership verification methods that align with the decentralized ideals of web3 (Gupta et al., 2022).

2.4.1   Real-World Examples of NFT Scams

In the nascent world of NFTs, scams exploit the excitement and trust within digital communities. A prevalent form of deceit involves Discord impersonation, where scammers masquerade as official admins or bots of NFT projects. They lure users with promises of exclusive minting opportunities or giveaways. The scam unfolds as users are directed to fraudulent links that steal their personal and financial information or trick them into sending cryptocurrency with the false hope of receiving NFTs. These incidents have led to substantial losses for individuals and have eroded trust in legitimate NFT projects, as users become wary of engaging with community offers and announcements (Ebutemetaverse, 2022).

Another notorious scam that duped eager collectors was the Fake Bored Ape Yacht Club Offer. In this scheme, fraudsters crafted a convincing copycat website of the well-known Bored Ape Yacht Club (Coindesk, 2021), offering sales of non-existent NFTs. Unsuspecting buyers, drawn by the prestige of the Bored Ape brand, were deceived into purchasing what they believed were rare digital assets. Upon transfer of the funds, the buyers received nothing in return, with losses amounting to significant sums given the high value of genuine Bored Ape NFTs. The scam not only caused financial damage to victims but also prompted a response from the real Bored Ape Yacht Club, warning their community about the risks of such scams and urging verification before any purchase.

2.5   IPFS and the Risks of Digital Storage in NFTs

The InterPlanetary File System (IPFS) represents a paradigm shift in the way we think about digital storage, particularly in the context of non-fungible tokens (NFTs). As a decentralized storage solution, IPFS allows NFT data to be stored across a global network of nodes, rather than on a single server. This architecture not only enhances the speed of access and redundancy but also fundamentally alters the reliance on traditional web hosting paradigms. By distributing data across numerous nodes, IPFS mitigates the risks associated with single points of failure, which is crucial for the NFT market where the provenance and availability of digital assets are paramount. The decentralization inherent in IPFS also introduces the potential for lower hosting costs and greater efficiency in data retrieval, offering a more scalable and resilient framework for NFT storage (Understanding how IPFS works (n.d.)).

However, the innovative approach of IPFS brings with it unique security concerns that must be addressed to ensure the integrity of NFT assets. One such concern is the risk of content permanence; if NFT data, which is often a tokenized digital file, is not properly pinned to reliable nodes, it may become inaccessible over time, undermining the value proposition of NFT ownership. Additionally, the open and distributed nature of IPFS means that content can potentially be exposed to malicious actors, posing a threat to the security of the stored NFTs. Ensuring that content is securely pinned and implementing measures to safeguard against the distribution of malicious content are vital steps in leveraging IPFS for NFT storage without compromising the security and longevity of the assets.

2.5.1 Path-Based vs. Subdomain-Based Gateways

In the architecture of IPFS gateways, two primary methods for accessing content can be observed: path-based and subdomain-based gateways. Path-based gateways serve content through various paths under a single domain, offering a straightforward approach to data retrieval. However, this approach bears security implications, notably an increased risk of cross-site scripting (XSS) attacks, as content from different origins is served under the same domain, potentially enabling malicious scripts to run and affect other data paths.

Conversely, subdomain-based gateways allocate a unique subdomain for each piece of content, thereby creating a more isolated environment for each data item. This method significantly mitigates the risk of XSS attacks by ensuring that content served from one subdomain cannot interfere with content from another. In the context of securing NFT assets, subdomain-based gateways are preferred as they provide an additional layer of security, leveraging the DNS hierarchy for content separation and reducing the possibility of cross-site vulnerabilities.

2.5.2 Securing IPFS Content

Ensuring the security of content stored on the InterPlanetary File System (IPFS) is critical, given the decentralized and persistent nature of its architecture. One of the primary means to secure content is through encryption. By encrypting files prior to uploading them to IPFS, users can protect sensitive information from unauthorized access. This process means that even if the data is distributed across multiple nodes, only individuals with the correct decryption key can access the true content. Additionally, the integration of smart contracts or the implementation of off-chain access control mechanisms can provide granular control over who is permitted to decrypt or view the content, thus enforcing strict access guidelines. To further ensure the persistence and availability of content, content pinning strategies can be employed. This involves using reputable pinning services or establishing private IPFS clusters, which ensures continuous hosting of the content and that only authorized nodes can serve and access the data, maintaining the integrity and availability of the content within the IPFS network.

2.6 Securing the NFT Ecosystem: Best Practices and Future Directions

To fortify the NFT ecosystem against security breaches, it is essential to engage in vigilant authorization and signature practices. Users must scrutinize every authorization request and signature prompt, confirming that the actions they approve correspond with their intentions. This cautionary step is crucial as scammers often devise schemes to manipulate users into unintentional approvals, leading to unauthorized access or transfer of assets. Moreover, the classification of wallets plays a pivotal role in asset security. For substantial holdings, hardware wallets offer superior protection by securing private keys offline, away from the reach of online threats. Utilizing trusted crypto wallets and marketplaces, along with enabling two-factor authentication (2FA), significantly enhances the security layers around NFT assets, thwarting unauthorized access and reinforcing trust in the system.

The digital landscape is fraught with deceptive links and emails, which are often the harbingers of phishing attacks. Users must exercise due diligence by verifying the legitimacy of sources and refraining from sharing sensitive information like seed phrases or private keys. Protecting one's seed phrase with the same rigor as one would protect a physical key to a vault is imperative; it should never be stored digitally or disclosed publicly. Staying current with software and firmware updates is equally important as these often contain crucial security enhancements. Keeping abreast of the latest security trends and threats within the NFT space empowers users with knowledge, allowing them to adapt and refine their security practices proactively.

In the realm of NFT investments, thorough research is the bulwark against scams and fraudulent schemes. Investigating a project's background, the credibility of team members, the feasibility of their roadmap, and the vibrancy of their community engagement can provide insightful indicators of the project's legitimacy. Independent reviews and security audits are invaluable resources that provide an objective assessment of a project's security posture. The collective effort of staying educated, practicing rigorous security measures, and fostering community collaboration is the cornerstone of a secure and thriving NFT marketplace that balances innovation with user protection and maintains the integrity of digital asset exchanges.

2.7 Conclusion

In conclusion, the future of NFT security hinges on the adoption of a holistic approach that blends cutting-edge technology with proactive community involvement and responsive regulatory frameworks. The security best practices detailed in this chapter are not just recommendations but necessities for the maturation of the NFT market. By fostering collaboration among stakeholders, staying informed about emerging threats, and committing to ongoing education and innovation, the NFT community can aim to create a secure and resilient environment that supports the growth and sustainability of this transformative space.

References

BBC News. (2021, March 12). Bored Ape NFT Collector Loses $2.2M in Phishing Scam. Retrieved from https://www.bbc.co.uk/news/technology-56362174

Blockgeeks. (n.d.). The DAO Hack: What You Need to Know. Blockgeeks. https://blockgeeks.com/guides/dao-hack/

Bloomberg. (2021, August 25). Axie Infinity: How game is turning pandemic jobless into crypto NFT traders. https://www.bloomberg.com/news/articles/2021-08-25/axie-infinity-how-game-is-turning-pandemic-jobless-into-crypto-nft-traders

Benson, J. (2021a, September 17). OpenSea Refunds Ethereum Users Who Lost NFTs in Inactive Listing Exploit. Decrypt. https://decrypt.co/91513/opensea-refunds-ethereum-users-lost-nfts-inactive-listing-exploit

Benson, J. (2021b, September 21). OpenSea NFT Market Swells to Record $5 Billion in Ethereum. Decrypt. https://decrypt.co/91748/opensea-record-5b-ethereum-nft-market-swells

CoinDesk. (2021, May 4). Bored Ape Yacht Club Warns of Discord Scam Attempt After NFT Boom. CoinDesk. https://www.coindesk.com/business/2021/05/04/bored-ape-yacht-club-warns-of-discord-scam-attempt-after-nft-boom/

CryptoBriefing. (n.d.). Bored Ape NFT Collector Loses $2.2M in Phishing Scam. Retrieved from https://cryptobriefing.com/bored-ape-nft-collector-loses-2-2m-in-phishing-scam/

Das, D., Bose, P., Ruaro, N., Kruegel, C., & Vigna, G. (2022, November). Understanding security issues in the NFT ecosystem. In Proceedings of the 2022 ACM SIGSAC Conference on Computer and Communications Security (pp. 667-681).

Dent, S. (2021, October 13). Security flaws at NFT marketplace OpenSea left users' crypto wallets open to attack. Engadget. Retrieved from https://www.engadget.com/2021-10-13-opensea-nft-marketplace-security-flaws.html

ELLIPTIC. (2022). NFTs and Financial Crime. Retrieved from https://www.elliptic.co/hubfs/NFT%20Report%202022.pdf

Ebutemetaverse. (2022, July 4). Common NFT Scams and How to Avoid Them. Ebutemetaverse. https://ebutemetaverse.com/common-nft-scams-and-how-to-avoid-them/

Geron, T. (2023, January 21). The NFT That Was a Frosties Ad. Protocol. https://www.protocol. com/fintech/frosties-nft-rug-pull

Gupta, Y., Kumar, J., & Reifers, D. A. (2022). Identifying security risks in NFT platforms. arXiv preprint arXiv:2204.01487. Retrieved from https://arxiv.org/abs/2204.01487

Hern, A. (2021, September 1). Collector Buys Fake Banksy NFT for $244,000. The Guardian. https:// www.theguardian.com/technology/2021/sep/01/collector-buys-fake-banksy-nft-for-244000

Introducing MarketMaker. (n.d.). Intertrust. Retrieved from https://www.intertrust.com/news/ introducing-marketmaker/

Kale, A. M., Yadav, A. K., & Singh, A. V. (2021, September 10). The Role of Artificial Intelligence and Big Data in the Fight Against COVID-19 Pandemic: A Comprehensive Review. PMC (PubMed Central). https://www.ncbi.nlm.nih.gov/pmc/articles/PMC9759957/

McCoy, K. (2014). Quantum. Retrieved from https://mccoyspace.com/project/125/

NFT collector loses $38,000 in what he believes is an OpenSea or Rarible glitch. (2022, January 1). Web3 Is Going Just Great. Retrieved from https://web3isgoinggreat.com?id=2022-01-01-3

NFTNow. (2021, November 16). Meet the 16-Year-Old Dev Who Became a Hero in the NFT Trader Exploit. NFTNow. https://nftnow.com/features/ meet-the-16-year-old-dev-who-became-a-hero-in-the-nft-trader-exploit/

OpenSea NFT marketplace hacked. (n.d.). Security Affairs. Retrieved from https://securityaffairs. com/128207/breaking-news/opensea-nft-marketplace-hacked.html

OpenZeppelin Security Audits. (n.d.). OpenZeppelin. Retrieved from https://openzeppelin.com/ security-audits/

Paganini, P. (2021, October 11). OpenSea NFT Marketplace Hacked. Security Affairs. https://secu-rityaffairs.co/wordpress/128207/cyber-crime/opensea-nft-marketplace-hacked.html

Snowden, E. [Foundation]. (n.d.). Stay Free. Retrieved from https://foundation.app/@Snowden/ foundation/24437

The rise of Axie Infinity and how a game is turning pandemic jobless into crypto traders. (2021, August 25). Bloomberg. Retrieved from https://www.bloomberg.com/news/articles/2021-08-25/ axie-infinity-how-game-is-turning-pandemic-jobless-into-crypto-nft-traders

The theft of NFTs and the crypto bear market. (2022). NFT Report. Retrieved from https://www. elliptic.co/hubfs/NFT%20Report%202022.pdf

Reuters. (2021, August 12). How Hackers Stole $613 Million in Crypto Tokens from Poly Network. Reuters. https://www.reuters.com/technology/ how-hackers-stole-613-million-crypto-tokens-poly-network-2021-08-12/

Understanding how IPFS works. (n.d.). IPFS Docs. Retrieved from https://docs.ipfs.tech/concepts/ how-ipfs-works/

Yirka, B. (2022, June 6). The NFT Insider Trading Case. Tech Xplore. https://techxplore.com/ news/2022-06-nft-insider-case.html

Crypto News. (2024, January 8). Smart Contract Vulnerabilities in NFTs. Retrieved from https:// crypto.news/smart-contract-vulnerabilities-in-nfts/

4irelabs. (2023, November 4). Top 17 Smart Contract Hacks. Retrieved from https://4irelabs.com/ articles/top-17-smart-contract-hacks/

Hacken. (2023, December 4). Smart Contract Vulnerabilities. Retrieved from https://hacken.io/ discover/smart-contract-vulnerabilities/

Lisa JY Tan 's work as the founder and lead economist at Economics Design has made her a pioneer in the design and engineering of digital ecosystems. With a track record of over 30 token economies and 50 token analyses, Lisa's work is characterized by a research-focused approach and a deep understanding of the potential of blockchain technology. As a highly sought-after speaker at conferences and forums worldwide, Lisa's expertise in token economics and DeFi has established her as a respected authority in the field of digital ecosystems.

Chapter 3
DAO Security

Carlo Parisi and Dmitriy Budorin

Abstract This chapter provides a comprehensive exploration of Decentralized Autonomous Organizations (DAOs), a groundbreaking innovation in blockchain technology and governance. It introduces DAOs as self-executing smart contracts, revolutionizing decision-making and resource allocation.

Through compelling case studies, the chapter showcases real-world examples of DAOs, highlighting their successes and challenges. It addresses the vulnerabilities and hacks that have plagued DAOs, emphasizing the need for robust security measures. Additionally, it sheds light on the shadowy aspects of centralized DAOs, emphasizing the importance of decentralization principles.

Looking toward the future, the chapter discusses potential advancements in DAO technology and governance, proposing innovative solutions to overcome current limitations. It contemplates the evolution of DAOs in shaping a more equitable and decentralized world.

3.1 Introduction

A Decentralized Autonomous Organization (DAO) represents an innovative approach for multiple individuals to collaborate without a single leader making all the decisions. It operates within the land of digital currencies and blockchain technology, utilizing smart contracts and blockchain records for transparent decision-making. In contrast to traditional hierarchical structures, DAOs distribute decision-making power among their members, all of whom hold tokens representing their influence. These tokens link members' financial interests to the collective choices made, creating a sense of responsibility and accountability.

C. Parisi (✉) · D. Budorin
Hacken, Lisbon, Portugal
e-mail: c.parisi@hacken.io; d.budorin@hacken.io

© The Author(s), under exclusive license to Springer Nature
Switzerland AG 2024
K. Huang et al. (eds.), *Web3 Applications Security and New Security Landscape*,
Future of Business and Finance, https://doi.org/10.1007/978-3-031-58002-4_3

DAOs essentially function as digital clubs where participants receive special tokens as membership credentials. These tokens serve as the basis for voting on various matters, from project selection to fund allocation. Importantly, DAOs rely on smart contracts, which are like automated digital rulebooks, guiding decision-making based on predefined conditions recorded on the blockchain.

One notable feature of DAOs is their commitment to decentralization. Unlike conventional organizations with a top-down leadership structure, DAOs empower a broad base of members to share decision-making authority. This inclusivity is similar to a democratic system, where each member has a voice.

Participation is another core principle of DAOs, as even individuals with relatively few tokens can actively contribute to the organization's direction. This encourages a sense of involvement and dedication among all members.

Transparency is an essential aspect of DAO operations. All decisions and actions are publicly recorded on the blockchain, creating a system of checks and balances within the community. This public accountability incentivizes members to act in the best interests of the collective.

Furthermore, DAOs have the remarkable ability to unite individuals from diverse geographical locations into a collaborative digital community. The internet serves as the platform through which token holders interact and collaborate on shared objectives.

However, like any innovation, DAOs come with their own set of challenges. Decision-making within DAOs can be time-consuming due to the need for widespread voting and coordination among participants. Ensuring that all members are well-informed about ongoing initiatives is another hurdle, as diverse backgrounds and levels of understanding can slow down the process. Inefficiency can also occur, with DAOs potentially getting stuck in lengthy discussions rather than swiftly implementing decisions. Finally, security remains a critical concern, demanding technical expertise to safeguard against vulnerabilities and threats.

In conclusion, Decentralized Autonomous Organizations are pioneering entities that redefine collaborative decision-making in the digital age. Their commitment to decentralization, participation, transparency, and global community-building make them a promising avenue for reshaping how we work together. While they grapple with certain challenges, the potential for DAOs to democratize governance and promote interconnectedness in the digital world is a testament to their significance in our evolving landscape.

3.2   The Different Types of DAOs

Decentralized Autonomous Organizations come in several flavors, each tailored to serve distinct purposes within the blockchain landscape. It's crucial for businesses to grasp the functions and objectives of these diverse DAO types before launching or participating in one to ensure alignment with their goals (Chamria, 2023).

3.2.1 Protocol DAOs

Protocol DAOs are specifically designed to oversee various functions within decentralized protocols, such as lending and borrowing. These DAOs help these platforms efficiently manage their operations and decision-making processes.

MakerDAO: Operating on the Ethereum blockchain, MakerDAO enables users to lend and borrow tokens with adjustable interest rates and flexible repayment terms. MKR governance token holders participate in decisions like collateral requirements for collateralized debt positions (CDPs), annual borrowing limits, and protocol shutdown plans in case of an Ethereum crash.

Uniswap: Uniswap's governance token, UNI, empowers the community to vote on the platform's development and operations. UNI holders govern Uniswap, manage the UNI community treasury funds, and control the protocol fee switch. Proposals require at least 25,000 affirmative UNI votes for consideration.

Yearn Finance: Yearn Finance employs its YFI governance token to allocate funding to DAO Vaults. Approved DAOs can receive funding from YFI token holders, bridging gaps in traditional HR and payroll systems. The creator, Andre Cronje, introduced Coordinape, a tool for allocating funds and rewarding contributors.

3.2.2 Grant DAOs

Grant DAOs are established to strategically deploy capital resources sourced from charitable donations across the web3 ecosystem. Their primary goal is to provide financial support to projects that drive meaningful change, promoting fairness and sustainability. Grant DAOs introduce transparent and democratic processes to revolutionize project funding, setting them apart from traditional organizations.

MolochDAO: Rooted in the Ethereum ecosystem, MolochDAO facilitates collective decision-making and resource allocation to support Ethereum's infrastructure development.

GitCoin DAO: GitCoin DAO supports thousands of projects in growing their open-source ecosystems and has distributed millions in funding.

3.2.3 Philanthropy DAOs

Philanthropy DAOs aim to advance social responsibility by uniting around a common mission to create a positive impact in the Web3 space. These decentralized communities harness blockchain technology's capabilities to bring about significant change.

Big Green DAO: The first nonprofit philanthropic DAO is affiliated with Big Green, a renowned charity. It focuses on educating people about sustainable food growth, nutrition, and mental health.

UkraineDAO: An example demonstrating the rapid impact potential of philanthropic DAOs. It quickly gathered over $3 million in ETH to support the Ukrainian Army, with contributions from PleasrDAO, Trippy Labs, and Russian art collective Pussy Riot.

3.2.4   Social DAOs

Social Decentralized Autonomous Organizations provide platforms for like-minded individuals, including builders, artists, and creatives, to collaborate, learn, engage in projects, and earn rewards. These DAOs, sometimes called Creator DAOs, often have entry barriers, such as token ownership or invitations.

Developer DAO: A community of web3 enthusiasts and developers working together to shape the future of web3.

3.2.5   Collector DAOs

Collector DAOs enable members to pool resources and collectively invest treasury funds in blue-chip NFT art and collectibles. This collective approach allows members to access investments that may have been out of reach individually.

ConstitutionDAO attempted to purchase a first-edition copy of the United States Constitution, raising $47 million worth of ETH in just 1 week.

3.2.6   Investment and Venture DAOs

Venture DAOs consolidate capital from various sources to fund early-stage web3 startups, protocols, and off-chain investments. They offer a more inclusive and accessible investment approach, allowing a wider range of individuals to participate in the investment process.

MetaCartel provides financial support and operational guidance to emerging decentralized applications (dApps).

BessemerDAO launched by Bessemer Venture Partners, this DAO shares insights and resources related to the crypto industry.

3.2.7 Media DAOs

Media DAOs redefine conventional media platforms by producing content directed by the community rather than corporate interests. These DAOs distribute profits among their members, empowering them to contribute to content creation and profit-sharing.

BanklessDAO promotes banking-free money systems through media, culture, and education.

Forefront: A token-permissioned media DAO that shares insights and community insights related to social tokens, NFTs, and other DAOs.

3.2.8 SubDAOs

SubDAOs are an innovative form of DAO focused on delegating specific responsibilities within a DAO to a select group of members. They enhance decentralization while allowing for specialization and increased operational efficiency.

Balancer Protocol Balancer implemented subDAOs to streamline decision-making and implementation, making the overall DAO more efficient.

3.3 Case Studies

3.3.1 Juno: Proposal 16, a Dangerous Precedent

Juno Network, a blockchain platform operating on the Cosmos network, found itself embroiled in a contentious decision known as Proposal 16. This proposal posed a critical question to the community of JUNO token holders: should a substantial portion of tokens held by one community member be confiscated, with the intention of either returning them to the community pool or permanently destroying them? The proposed amount to be reclaimed was a staggering 3,103,947 JUNO tokens, with a market value of approximately $117 million at the time of the vote.

The driving force behind this proposal was an allegation that the targeted wallet had engaged in manipulative tactics during a recent airdrop, resulting in the accumulation of an excessive number of JUNO tokens. These tokens, like many DeFi assets, granted significant voting power within the Juno Network.

The failure to pass this proposal carried multiple risks, as outlined by the proposer. Firstly, the existence of a single wallet with control over "half of the quorum" necessary for votes raised significant concerns about centralization. Secondly, with control over such a substantial token supply, the holder had the potential to swiftly disrupt DEX liquidity, posing a serious threat to various crypto markets trading the JUNO token.

Fig. 3.1 Juno Community voting results

Additionally, the holder's sizable token holdings gave them considerable influence over validators, potentially incentivizing fraudulent behavior within the network. Lastly, the existing situation was causing widespread apprehension and fear within the Juno community, adding weight to the proposer's arguments.

These compelling arguments swayed a considerable portion of the Juno community, leading to a narrow victory for the "yes" votes during the subsequent vote (Fig. 3.1).

This decision ignited passionate and divided responses within the broader crypto community. Some likened the entity in question to a "ticking time bomb," while others expressed reservations about establishing a "dangerous precedent."

Before the vote on March 14, 2023, the entity in question issued a statement in a now-deleted Medium post, pledging to return all assets to users should Proposal 16 be rejected. However, they were not given the opportunity to fulfill this promise.

In the interim, the crypto community, as well as media outlets, will continue to dissect and evaluate the far-reaching implications of this decision. This event has provoked critical discussions about the core principles of decentralization, community governance, and the immutability of blockchain code in the crypto world (Kelly, 2022).

3.3.2   STEEM: Hostile Takeover

A crypto scandal has erupted in the wake of TRON CEO Justin Sun's alleged manipulation of on-chain governance within the Steem blockchain.

Steem operates on a Delegated Proof-of-Stake (DPoS) consensus mechanism, where 21 delegates, also known as witnesses, are democratically elected by Steem Power holders to validate transactions and create blocks.

The turmoil began when Steem blockchain witnesses initiated a soft fork to diminish the influence of the TRON Foundation, resulting in the removal of a substantial number of STEEM tokens from their control. In response, TRON leveraged major exchanges to regain control over the platform, a move that has drawn widespread criticism within the crypto community for its centralization of power.

Justin Sun's involvement in Steem started with the acquisition of Steemit Inc. and the announcement of a strategic partnership between TRON and Steem. However, discontent emerged among the Steem community members, leading to threats of a soft fork by witnesses. This prompted Sun to take a more assertive stance.

The majority of Steem's witnesses were subsequently replaced by accounts believed to be connected to TRON, with major crypto exchanges like Binance, Huobi, and Poloniex exercising their substantial voting power by delegating votes to Sun-controlled accounts.

Sun's acquisition of Steemit Inc. also granted access to a significant quantity of STEEM tokens obtained through a pre-mining operation referred to as "ninja-mined stake." Initially, witnesses voted for a soft fork to prevent Sun from using these tokens for governance, as the former Steem founder, Ned Scott, had promised that they would not be used for voting. However, Sun's takeover changed this.

TRON's official statement acknowledged the introduction of Steem Witnesses Soft Fork 22.2, which aimed to freeze the original Steemit shares. TRON intended to use these tokens to enhance the Steem blockchain's development but was over-powered by the soft fork. Consequently, TRON took action to reclaim the stake and appoint new witnesses to promote a healthier ecosystem.

Critics argue that Soft Fork 22.2 was structured maliciously, targeting specific accounts and depriving them of their rights and asset ownership, potentially consti-tuting an illegal act. This move also raised concerns about the threat it posed to decentralization and the core values of the Steem blockchain community.

In response to community backlash, Justin Sun presumed that their intention was not to control the STEEM network but to protect Steemit's stake and STEEM hold-ers' interests. Sun pledged to withdraw the votes once they were certain that "mali-cious hackers" could no longer harm STEEM, returning voting rights to the community and withdrawing votes from all exchanges involved in the takeover.

The crypto community's reaction to the takeover has been harsh, with Sun's actions drawing condemnation for undermining decentralization and community values. Major exchanges like Binance have faced criticism for their involvement in the takeover, leading to the likely withdrawal of their votes in response to commu-nity feedback. (Simmons, 2020).

3.3.3 Mango Markets: "A Highly Profitable Trading Strategy"

Mango Markets has reported a hack totaling approximately $112 million in digital assets. The attacker utilized a method known as oracle price manipulation, a form of economic attack that has previously impacted other DeFi protocols.

The attacker successfully withdrew various digital assets, including $53.7 mil-lion in USD Coin (USDC), $3.2 million in tether (USDT), and solana (SOL). In an unusual turn of events, the hacker has proposed returning a portion of the stolen funds, which includes Marinade-staked solana (MSOL), native SOL, and Mango's MNGO governance token. The remainder is labeled as a "bounty" to be kept by the attacker, pending approval from Mango's DAO community. An interesting fact to note is that most of the votes for the approval of the bounty came from the hacker themself.

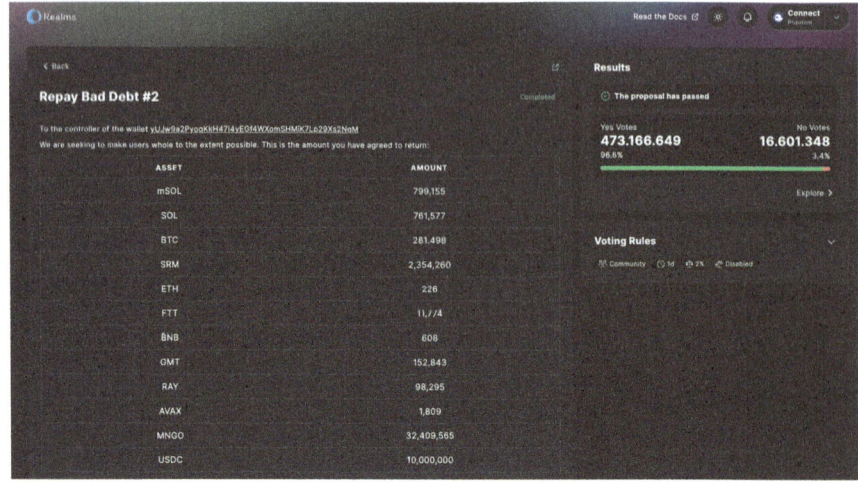

Fig. 3.2 Attacker keeps a substantial portion of what was stolen

To execute this proposal, the attacker is suggesting that Mango use its treasury's 70 million USDC to repay the "bad debt." The attacker also requests that Mango's DAO abstains from pursuing criminal investigations or freezing the attacker's funds once the crypto assets are returned.

However, not all the assets will be returned, and the attacker is seemingly requesting to keep a substantial portion of what was stolen, which deviates from typical practices seen in "white hat" hacker or bug bounty scenarios. (Fig. 3.2).

The situation has prompted Mango to request users to refrain from depositing assets until further clarity is achieved. The attack leveraged price oracle manipulation, where corrupted data feeds allowed unintended transactions. The manipulation occurred as the attacker manipulated their collateral value through the platform before taking out substantial loans from Mango's treasury (Blockworks, 2023).

3.3.4   Beanstalk: A Very Expensive Flashloan

In April 2022, Beanstalk, a permissionless fiat stablecoin protocol, was exploited in a massive hack worth $181 million. The attacker used flashloans from Aave, Uniswap, and SushiSwap to manipulate the governance system. By submitting malicious proposals and exploiting the lack of execution delay, the attacker gained enough voting power to execute the proposals and make a significant profit. Beanstalk is an Ethereum-based protocol that aims to maintain the value of its native stablecoin, Bean, at $1.

The vulnerability in Beanstalk's smart contract was due to the existence of the "emergencyCommit" function in the "GovernanceFacet" implementation contract.

This function allowed the attacker to execute active proposals with a 2/3 voting percentage. Flashloans were used to acquire the necessary voting power, enabling the attacker to execute the malicious proposal and transfer funds to their contract.

A simplified proof of concept (PoC) demonstrated the attack process, swapping ETH for BEAN tokens, proposing a malicious proposal (Bip18), flashloaning stablecoins, depositing them into Beanstalk, leveraging the emergency commit function to execute the malicious proposal, and swapping tokens back to stablecoins to repay the flashloan and make a profit.

The Beanstalk governance attack highlights the importance of secure pattern implementation in DeFi protocols. The ability to vote and execute proposals in the same transaction can leave a protocol vulnerable to governance attacks, especially when flashloans are involved. Beanstalk responded to the hack by replacing the GovernanceFacet implementation with a community-run multisig wallet (Immunefi, 2023).

3.3.5 Tribe DAO: What Not to Do in Case of a DAO Shutdown

Tribe DAO, the entity behind the FEI stablecoin and Rari Capital Markets, faced a shutdown due to challenges in the macro environment, regulatory concerns, and product-market fit issues amid the bear market. The shutdown proposal created significant drama within the community.

Initially, there was controversy over the allocation of funds for repayment. It was suggested that most of the redemption money would not go to compensate victims of the Rari hack but would be retained by the team. Prominent figures like FRAX founder Sam Kazemian argued for full reimbursement to hack victims, emphasizing that DAOs like Olympus and FRAX had provided liquidity support to Rari markets early on.

After passionate debates, the community ultimately voted in favor of fully repaying the hack victims.

The next challenge was selling off the protocol's assets, which included various volatile crypto assets like stETH and governance tokens. Selling them on-chain would have caused significant price impact and reduced redemption value for TRIBE holders. Therefore, the decision was made to conduct off-chain over-the-counter (OTC) sales.

Several players in the crypto space, including protocols and market makers, submitted bids for these assets in OTC deals.

As a result of these sales, sufficient funds were raised to cover the entire redemption of FEI into DAI, allowing the redemption process to begin.

The Tribe DAO wind-down highlighted the need for more efficient mechanisms for handling DAO wind-downs and OTC sales of crypto assets. It also emphasized the importance of a decentralized exchange (DEX) tailored for OTC swaps and OTC trading tools within the DeFi ecosystem (xWailord, 2023).

3.3.6 Temple DAO: Smart Contracts Security Should Be a Must

On October 11, 2022, Temple DAO, a decentralized finance (DeFi) protocol, suffered a hack resulting in a loss of over $2.3 million. The hack was initially identified by Twitter user Spreekaway and later confirmed by blockchain analytics company Peckshield.

The hacker initiated the attack through SimpleSwap and subsequently transferred 1831 ETH to a new address, 0x2B63d. The hack involved the theft of 321,154 xLP tokens from the xLP Staking contract, which were then converted into 1,418,303 $TEMPLE tokens and 1,262,438 $FRAX tokens. Subsequently, the TEMPLE tokens were exchanged for FRAX.

The exploit was made possible by the hacker taking advantage of a vulnerability in the StaxLPStaking contract.

To prevent further accidental usage, Temple DAO temporarily took down its decentralized application. The project's team has also called on the hacker to return the stolen funds and has offered a legal bounty in return (Adejumo, 2022).

3.3.7 The DAO: The Code Is Law, Until Is Not

In 2016, "The DAO" was created as a decentralized investment fund, raising $150 million worth of ether through its community token.

However, less than 3 months after its launch, the DAO fell victim to a "blackhat" hacker who used a "reentrancy" attack to drain most of the $150 million in ETH from the smart contract. This event raised questions about Ethereum's credibility and led to deep ideological debates within the community.

One viewpoint emphasized the immutability and tamper-resistance of blockchain technology, advocating for non-intervention even in the face of serious consequences. On the other hand, the theft of people's savings and the damage to public confidence in blockchain technology called for intervention to protect investors and prevent theft.

During these debates, a "whitehat" hacker group intervened using the same reentrancy exploit to rescue funds and refund them to investors. Meanwhile, the Ethereum core team faced a challenging decision. To stop the hacker, they could fork the Ethereum blockchain. This decision contradicted Ethereum's principles, because it appeared to be a centralized action.

Despite resistance from some miners, 85% of the community voted in favor of the fork, resulting in two parallel Ethereum blockchains: Ethereum Classic and the current Ethereum chain. This historical event significantly impacted the cryptocurrency space (Pratap, 2022).

3.4   Challenges and Opportunities

3.4.1   Challenges

While DAOs offer opportunities for effective coordination in decentralized environments, they face substantial challenges. These challenges include infrastructure development, legal complexities, blockchain vulnerabilities, governance effectiveness, and questions about community inclusivity versus profit-driven motives. Investing in or establishing a DAO is not a straightforward process, but navigating these challenges can be facilitated by a dedicated team of experts.

3.4.1.1   Inadequate Infrastructure

Given their purely digital and decentralized nature, DAOs often lack the well-established and efficient infrastructure enjoyed by their traditional counterparts. Many DAOs must develop their own tooling and infrastructure before launching, and these solutions can be somewhat unconventional.

DAOs face challenges in setting up essential infrastructures for reporting, treasury management, governance, payroll, communication, identity management, and more. Without proper infrastructure for these functions, DAOs risk losing members and struggling to gain traction. However, ongoing innovations and collaborations within the Web3 space are gradually addressing some of these infrastructure issues.

3.4.1.2   Legal Complexities

DAOs share similarities with various modern organizational structures like corporations, non-profits, partnerships, and cooperatives. However, these traditional entities are governed by the legal frameworks of specific jurisdictions.

In contrast, DAOs are decentralized and lack a fixed location, making it challenging to fit them within existing legal frameworks. They grapple with regulatory issues such as taxation of DAO tokens, treasury management, insurance for investments, AML and CFT policies, and liability for their actions. The absence of a clear legal system to address these challenges limits DAOs' ability to establish a foothold in the physical world.

3.4.1.3   Navigating Blockchain Risks

While blockchains themselves are highly secure, the smart contracts running on them may have vulnerabilities. Several smart contract networks have suffered attacks, resulting in the loss of significant funds.

3.4.1.4 Effective Governance

Many DAOs employ a democratic one-token-one-vote system, which, while inclusive, may not be suitable for managing complex decisions requiring expertise. In traditional organizations, hierarchies exist for efficient decision-making, as not all members can effectively vote on every matter.

Experienced individuals are essential for making technical decisions, but some DAOs struggle to accommodate this need. Some DAOs, like the ENS DAO, explore hybrid governance structures where members vote in qualified personnel to handle critical decisions transparently. However, this approach comprises the complete decentralization ideal, inviting criticism from Web3 enthusiasts.

3.4.1.5 Addressing Inactive Token Holders

Inactive token holders can disrupt DAO functionality. While vote delegation is a solution, it introduces centralization concerns. Some DAOs incentivize active token holders with rewards while penalizing inactive ones, attempting to strike a balance between participation and centralization.

3.4.1.6 Community Over Profit

Critics argue that many DAOs primarily benefit wealthy individuals, contrary to the decentralization and community empowerment goals they profess.

Additionally, some DAOs receive funding from venture capitalists (VCs), contradicting their intended decentralized and community-driven nature. VCs are increasingly investing in DAOs, combining their operational expertise with community-driven principles.

3.4.2 Opportunities

DAOs offer a host of benefits, including enhanced decentralization, "skin in the game" for participants, and the ability to improve community-driven initiatives. As blockchain technology continues to evolve, these advantages are likely to become even more pronounced, making DAOs a crucial element in the ongoing development of Web3.

3.4.2.1 Autonomous Structure

One of the foremost benefits of DAOs lies in their very definition—an autonomous structure. Unlike traditional organizations burdened by top-down hierarchies, DAOs empower each member with autonomy. This autonomy translates into freedom from centralized control, facilitating more efficient decision-making and operations.

3.4.2.2 Equal Stakes

In discussions about the advantages of decentralized autonomous organizations, the focus naturally shifts to inclusivity. In DAOs, every investor has the opportunity to influence the organization's future. The number of DAO tokens owned by an investor determines their voting power within the organization. Furthermore, any stakeholder can propose ideas or improvements, which are visible to all members. This transparency extends to decentralized governance, where every member can express their opinions on proposed changes.

3.4.2.3 Neutrality

Conflict and disputes in traditional organizations often arise from discrepancies in established rules. A key advantage of DAOs is the neutral environment they encourage. With no intermediaries or overbearing managers, power struggles are minimized.

3.4.2.4 Accountability

Transparency is a central element of DAOs, and with it comes accountability. Every member of the organization is accountable for their actions. For instance, when submitting proposals or ideas to the DAO, members must consider the associated costs. Evaluating proposals requires contributions of resources from every member, incentivizing careful refinement of ideas to steer the organization in the right direction.

The importance of accountability in DAOs also extends to the use of blockchain technology. All activities within the organization are recorded on the blockchain, creating an immutable ledger that allows for transparent tracking of transactions. This transparency not only reduces the risk of scams, such as rug pulls, but also instills confidence among investors, as evidenced by the growing popularity of DAO tokens in the market.

3.4.2.5 "Skin in the Game" for Participants

An integral aspect contributing to the success of DAOs is the concept of "skin in the game" for every participant.

DAOs that employ governance tokens require users to expend or commit their tokens to cast votes. This incentivizes thoughtful consideration for each vote, as governance tokens grant users the ability to influence the DAO's future.

3.4.2.6 Community-Driven

DAOs offer the added advantage of enabling global communities to easily connect and collaborate on shared visions. With an internet connection and governance tokens, virtually anyone can contribute to shaping the future of Web3. Whether it involves influencing the direction of domain management, as seen with ENS, or creating a popular play-to-earn cryptocurrency, DAOs extend accessibility to individuals who may have been excluded in the past.

Additionally, participating in a DAO promotes a sense of ownership, similar to owning stock in a company. This ownership mentality not only fuels innovation but also opens the door to potential financial rewards.

3.5 Attacks and Defense Measures

3.5.1 Attacks

3.5.1.1 Smart Contract Vulnerabilities

One of the primary security concerns in DAOs stems from the smart contract code itself. Flaws or vulnerabilities in the code can lead to glitches and, worse, exploitable weaknesses. Attackers, upon discovering a loophole in the code, can potentially dismantle a DAO's structure or disrupt its tokenomics. This places immense pressure on developers to ensure that smart contracts are flawless before deployment to prevent the rapid destruction of an entire project.

An example of such a risk materialized with Temple DAO, where a vulnerability allowed an attacker to manipulate staking contracts and drain $2.3 million from the DAO. This incident illustrates the ease with which problems can slip past scrutiny.

3.5.1.2 Governance Risks

The nature of DAO governance also poses a significant area of concern. Many DAOs still rely solely on token-weighted voting, where those with substantial token holdings wield disproportionate influence over the project's future. This can lead to centralization, where entities collude to steer the DAO away from the community's best interests.

Governance attacks can also occur stealthily over an extended period, as attackers create numerous anonymous accounts to gradually accumulate governance tokens while maintaining the façade of ordinary holders. These dormant accounts often go unnoticed for prolonged periods, given the low voter participation typical in many DAOs. Over time, these Sybil wallets can amass sufficient power to unilaterally control governance without the community's ability to respond.

Market mechanisms for token allocation are incapable of distinguishing between users aiming to contribute positively to a project and those seeking to disrupt or control it. Both groups appear behaviorally indistinguishable in a public market-place, as both are willing to acquire tokens at escalating prices (a16z, 2022).

3.5.1.3 Response Time and Authority

The decentralized nature of DAOs can lead to delayed response times in the event of an attack. Even if an attack is detected promptly, major decisions, such as halting transactions, require community votes. This process can take hours, providing attackers with ample time to inflict substantial damage. Imagine a bank being robbed in real time, but security measures cannot be enacted until a board of directors conducts a vote.

3.5.2 Defense Measures

To mitigate these security risks, DAOs must implement a comprehensive system of checks and balances:

Quality Control and Audits Smart contract code must undergo thorough, independent audits before deployment. Regular audits are crucial, especially if the code is evolving.

Real-Time Network Monitoring implements publicly available tools for real-time monitoring of the entire ecosystem. This can help identify unusual transactions or metrics as they occur.

Community Communication maintains transparent communication channels with the community to disseminate critical information promptly. Engage the larger user base in both system development and feedback.

To improve the security of DAO protocols, designers must skillfully navigate the trade-offs between open decentralization of governance and protection against attackers exploiting governance mechanisms. A framework for assessing vulnerability can be concisely expressed in an equation:

For a protocol to be considered secure against governance attacks, an attacker's profit should be negative. This equation serves as a guiding beacon for evaluating design choices when crafting governance rules. Designers can mitigate incentives for protocol exploitation through three strategic approaches:

Decreasing the Value of Attacks restricts the scope of governance to narrow the potential attack surface. Governance may begin with broader powers but gradually introduce friction as the project matures, requiring larger quorums for significant decisions.

Increasing the Cost of Acquiring Voting Power diminishes token liquidity to heighten the challenge for attackers in accumulating voting power. Encourage token holders to stake or introduce additional value beyond governance to align their interests with the project's success.

Increasing the Cost of Executing Attacks implements frictions that disincentivize attackers from exercising voting power, such as user authentication or time locks on tokens. Delegation and veto powers can also dissuade attackers and provide the community with time to respond to malicious proposals.

By implementing these strategies, DAOs can fortify their protocols against governance attacks, reducing the risk of disruption and encouraging a more resilient decentralized ecosystem.

Security in DAOs is a complex challenge that continues to evolve. There is no one-size-fits-all solution, and the landscape is ever-changing. However, by adopting the latest practices and encouraging a culture of vigilance, DAOs can significantly enhance their security posture and reduce the occurrence of substantial losses.

3.6   (De)Centralized Autonomous Organizations

While DAOs were initially envisioned as bastions of democracy, equality, and inclusivity, the present reality falls significantly short of these ideals in some cases.

In 2022, decentralized autonomous organizations gained immense popularity, particularly appealing in a world marked by growing inequality. The allure lay in their promise of more open and inclusive decision-making mechanisms. When functioning optimally, DAOs eliminate centralized hierarchies and encourage collective decision-making through on-chain voting.

Notable success stories include initiatives like granting UNI token holders the right to influence the operations of major crypto exchange Uniswap and the Ukraine DAO, which directs donations to support war-affected communities in Ukraine based on community input. However, DAOs still grapple with several challenges, including oversimplified structural designs, privacy concerns, and power imbalances associated with their voting procedures. To fully realize their potential in creating nuanced and decentralized power structures, DAOs must prioritize composable privacy and the decentralization of their formation processes.

Despite their reputation for being decentralized and autonomous, many DAOs are effectively controlled by a small group of major shareholders or developers with strong internal connections. A June 2022 report by Chainalysis examining ten prominent DAO projects found that, on average, less than 1% of all token holders wielded 90% of the voting power. Additionally, only 1 in 10,000 governance token holders had sufficient tokens to propose changes, and merely 1 in 30,000 could pass a proposal.

One of the reasons behind the default centralization of many DAOs, despite their decentralized designs, is that their formation predominantly occurs off-chain. Participants come together based on shared interests or personal connections, creating a secret hierarchy that affects their influence.

The loose and flat structure of DAOs allows power misappropriation by large token holders to go unchecked, a phenomenon referred to as the "tyranny of structurelessness." DAOs have attempted to formalize their structurelessness but have struggled to demonstrate its benefits.

While DAOs aim to enhance egalitarian decision-making through voting processes, participation is often low. Some abstain due to disinterest or to sabotage proposals, while requiring more members to vote can lead to a high failure rate due to poor engagement. Striking the right balance poses a challenge.

Many participants in governance token-based DAOs become passive, engaging primarily to acquire and trade tokens. An illustration of this occurred when a contentious vote resulted in Brantly Millegan's reinstatement to the board of the Ethereum Name Service Foundation. Millegan held a significant number of tokens and used them to tip the scales in his favor, as many users abstained from voting.

DAO voting suffers from a pronounced privacy issue, as voting outcomes are often entirely transparent. This transparency can hinder meaningful decision-making when everyone can see each member's stance via on-chain data. Embracing composable privacy, with flexible privacy mechanisms, can address this concern by allowing developers to decide what information remains private and what is publicly auditable.

To further improve the integrity of DAOs, embracing composable zero-knowledge proofs, which enable verification of knowledge without revealing it, especially in anonymous voting systems, will be crucial. Addressing the limitations of coin voting is another challenge that DAO creators must confront as these organizations continue to evolve in the coming years (Yin, 2023).

3.7 The Future of DAOs

The convergence of open-source software, blockchain technology, economic incentives, and smart contracts offers enhanced transparency, trust, adaptability, and speed. However, traditional corporate structures, driven by profit maximization and centralized control, have remained largely unchanged. DAOs represent a departure from this model, emphasizing transparency, global access, democratized decision-making, flexibility, and autonomous governance.

Today, DAOs are diversifying into various sectors, including venture funds, grant programs, social networks, video games, financial platforms, philanthropy, decentralized exchanges, and even meme-buying groups. These real businesses manage substantial capital, offer services to millions, and create innovative income opportunities.

DAOs offer potential for greater transparency, trust, adaptability, and speed but also face challenges in scalability, voter engagement, power concentration, cybersecurity, and privacy, limiting their adoption.

DAOs may redefine the future of work, workforce, and organizations:

Employee Engagement DAOs offer employees flexibility to choose projects aligned with their values and strengths, potentially stimulating greater fulfillment at work.

Voice in Governance Robust governance mechanisms in DAOs empowers token-holding members to participate in key decisions.

Work-to-Earn Models DAOs could shift the work landscape to a more flexible and decentralized structure, expanding opportunities for token holders, bounty hunters, and contributors.

These models will require further refinement but promise a more dynamic and accessible workforce.

DAOs face challenges like regulatory ambiguity, low voter engagement, and security vulnerabilities. Regulatory clarity, community development, and due diligence are essential to address these issues.

While DAOs may not replace large corporations, they hold promise in various use cases, such as decentralized exchanges, lending protocols, venture funds, and charitable trusts, offering diverse income opportunities for members.

The crypto industry is at a critical juncture, with potential for growth amid global economic challenges. DAOs represent an untapped opportunity, but realizing their potential requires robust regulation and a matured ecosystem. While challenges exist, the future holds promise for DAOs to reshape work, governance, and collaboration, provided we navigate this evolving landscape with determination and resilience (Das, n.d.).

3.8   Conclusion

Decentralized Autonomous Organizations represent a pioneering innovation that redefines collaborative governance and decision-making. As explored throughout this chapter, DAOs leverage the capabilities of blockchain technology and smart contracts to facilitate transparent, inclusive, and democratic participation within decentralized digital communities.

However, as highlighted through the compelling case studies, DAOs continue to grapple with monumental challenges that threaten their stability, security, and core values. Manipulation of governance systems, flash loan attacks, smart contract vulnerabilities, inactive token holders, centralization risks, and regulatory ambiguity have all emerged as pressing issues. Nevertheless, ongoing research and innovation aim to fortify DAO protocols through rigorous auditing, enhanced tokenomics, refined governance rules, composable privacy, and regulatory clarity.

The future offers immense promise for DAOs to revolutionize work, collaboration, and resource allocation on a global scale. By pooling together capital, ideas, and talent, DAOs can fund impactful projects, support charitable causes, and enable broader participation in investment decisions. As blockchain interoperability expands through bridges across networks, the possibilities for cross-chain DAOs will amplify.

However, realizing this potential requires overcoming limitations in scalability, security, voter engagement, and social inclusion. Crafting intuitive user experiences, fostering community spirit, and embracing diverse perspectives will be vital for DAOs to flourish. While the road ahead remains filled with uncertainty, the expanding ecosystem of researchers, builders, and visionaries seem poised to transform DAOs into inclusive and resilient forms of digital commonwealths. By upholding commitments to transparency, accountability, and collective empowerment, DAOs can progressively decentralize decision-making and distribute wealth to usher in a more equitable paradigm for cooperation at scale.

References

Adejumo, O. (2022, October 11). Temple DAO hacked for over $2.3M. CryptoSlate. https://cryptoslate.com/temple-dao-hacked-for-over-2-3m/

Chamria, R. (2023). A complete guide to different kinds of DAOs. Blockchain Deployment and Management Platform | Zeeve. https://www.zeeve.io/blog/a-complete-guide-to-different-kinds-of-daos/

DAO governance attacks, and how to avoid them - a16z crypto. (2022, July 28). A16z Crypto. https://a16zcrypto.com/posts/article/dao-governance-attacks-and-how-to-avoid-them/

Pradeep Mohan Das (n.d.). Envisioning the Future with Decentralized Autonomous Organizations (DAOs). www.linkedin.com. https://www.linkedin.com/pulse/envisioning-future-decentralized-autonomous-daos-pradeep-mohan-das/

Immunefi. (2023, January 19). Hack Analysis: Beanstalk governance attack, April 2022. Medium. https://medium.com/immunefi/hack-analysis-beanstalk-governance-attack-april-2022-f42788fc821e

Kelly, L. J. (2022, March 19). How The Juno Network DAO Voted to Revoke a Whale's Tokens. Decrypt. https://decrypt.co/95435/juno-network-dao-proposal-16-voted-to-revoke-tokens-from-whale

Mango markets mangled by oracle manipulation for $112M. (2023, September 21). Blockworks. https://blockworks.co/news/mango-markets-mangled-by-oracle-manipulation-for-112m

Pratap, Z. (2022). Reentrancy Attacks and the DAO Hack. Chainlink Blog. https://blog.chain.link/reentrancy-attacks-and-the-dao-hack/.

Simmons, J. (2020). Scandal? TRON takes over Steem blockchain by force Crypto News Flash. https://www.crypto-news-flash.com/scandal-tron-takes-over-steem-blockchain-by-force/.

xWailord. (2023). Tribe DAO shutdown drama and case for OTC trading in DEFI. Integral. https://integral.link/tribe-dao-shutdown-drama-otc-trading-defi/.

Yin, A. S. (2023, March 3). Why DAO voting is riddled with problems and voting is a farce. Forkast. https://forkast.news/why-dao-voting-is-problematic/

Carlo Parisi is a senior Solidity smart contract developer, senior auditor, and content creator with a degree in Computer Science. He has a deep knowledge in developing and auditing Solidity code, the main language used in Ethereum for smart contracts. He has been a Bitcoin Enthusiast since 2013, DeFi user since 2018.

Dmitriy Budorin is Founder & CEO at Hacken and Founder at HackenProof.
Dyma is a cybersecurity expert and crypto economy influencer with 14+ years of managerial expertise in cybersecurity as well as risks and controls audits. Dyma holds a master's degree in

International Economics and an MBA from the Kyiv Institute of Investment Management. He is a certified member of the Association of Chartered Certified Accountants (ACCA).

In 2017 Dyma established Hacken, a cybersecurity consulting firm. Five years later, Hacken is trusted by the largest crypto projects; the company's portfolio includes HackenAI, HackenProof, CER, and a suite of accompanying blockchain services. Dyma's effective leadership is what transformed Hacken from a startup into a major player in Web3 cybersecurity. The story of success is only gaining momentum.

As the company's Co-Founder and CEO, Dyma is responsible for leading the team of 100+ talented specialists and providing a vision of the future. Dyma consults the Ukrainian government on the adoption of a virtual economy. He is a regular participant in major Web3 cybersecurity events as an invited speaker.

Chapter 4
Crypto Asset Exchange Security

Carlo Parisi and Dmitriy Budorin

Abstract The rapid evolution of the crypto market, combined with the recent breach at FTX, has very clearly highlighted the importance of maintaining robust security measures for crypto asset exchanges. With a substantial amount of wealth and trust on the line, exchanges must be proactive, leveraging best practices and innovative technologies to safeguard their platforms and user assets.

4.1 Introduction

In this chapter, various methods for enhancing the security of centralized exchanges will be examined. The utilization of proof of reserves, particularly through techniques such as Merkle trees and zkSNARKs, will be discussed to establish the solvency of an exchange. Additionally, attention will be given to best practices for key management in companies and the judicious handling of exchange tokens, particularly in the context of considering them as collateral in lending ventures.

While these topics are crucial, there are many facets to exchange security that won't be exhaustively explored in this chapter, such as utilizing AI/ML for detecting fraudulent transactions, safeguarding your API with an API Gateway, and the importance of adopting DevSecOps.

Given the dual presence of centralized exchanges in both web2 and web3 realms, presenting a comprehensive list of vulnerabilities and potential attack vectors within this chapter's constraints is challenging. Therefore, after addressing the prevalent issues in web3, we will transition to case studies. These will spotlight exchanges that have either been involved in fraudulent activities or fell victim to cyberattacks, offering insights into their unique situations.

C. Parisi (✉) · D. Budorin
Hacken, Lisbon, Portugal
e-mail: c.parisi@hacken.io; d.budorin@hacken.io

© The Author(s), under exclusive license to Springer Nature
Switzerland AG 2024
K. Huang et al. (eds.), *Web3 Applications Security and New Security Landscape*,
Future of Business and Finance, https://doi.org/10.1007/978-3-031-58002-4_4

55

4.2   Proof-of-Reserve Techniques

Proof-of-reserve audits confirm that crypto exchanges indeed possess the assets they assert, thereby increasing investor trust in asset safety, especially following notable frauds that caused significant losses. These audits promote openness and prevent suspicious activities, vital for the industry's trustworthiness and expansion. Yet, many misunderstand proof of reserve. The top three verification methods are third-party audits, Merkle tree technology, and zkSNARK technology.

In the following sections, we'll delve deeper into Merkle trees and zkSNARK.

A key aspect to consider regarding proof of reserves is the inclusion of proof of liabilities. Demonstrating the exchange's solvency or as a separate proof, alone, proof of reserve reveals the exchange's funds. Without knowing the owed amount to clients, one can't ascertain the exchange's solvency.

Proof of reserve is essential in the cryptocurrency domain because it lets customers verify their asset safety, encourages exchange transparency, and ensures assets aren't lent out like traditional banks. Public disclosures, as seen with exchanges like Binance and KuCoin, offer transparency and security but aren't without flaws. They might show reserves only at a specific moment, and funds can be manipulated. Combining proof of reserve with other transparency tools, like live audits, can offer a clearer financial picture. While proof of reserve has its limitations, it's a step toward more transparency in the crypto industry. It's not just about trusting exchanges but also testing their claims. For the ultimate trust, one might consider withdrawing their crypto for personal safekeeping. Testing an exchange's resilience through stress tests, like rapid withdrawals, can also be a good gauge of its reliability.

The simplest method for a proof of reserve would involve publicly declaring users' holdings on a platform and letting them verify the accuracy of those declarations. However, this straightforward approach risks privacy breaches. As a result, more sophisticated techniques, such as zero-knowledge proofs and Merkle trees, are employed to safeguard user privacy while ensuring transparency.

4.2.1   Merkle Tree

Merkle trees allow for the validation of vast amounts of data without recalculating the entire dataset. They uniquely separate data from its "proof."

If the Merkle tree's hashes remain consistent, they are resistant to tampering. This means users can confirm the truth of a dataset using only a fragment of its data. The fundamental premise behind proof of reserve is assuring the public that their stored cryptocurrency matches their actual balance.

Picture leaf nodes representing user balances and the Merkle root as the aggregate of all balances in real-time (Fig. 4.1). If the exchange states its reserves, auditors can contrast what's owed against the exchange's assertion. This Merkle tree mechanism lets users validate these claims.

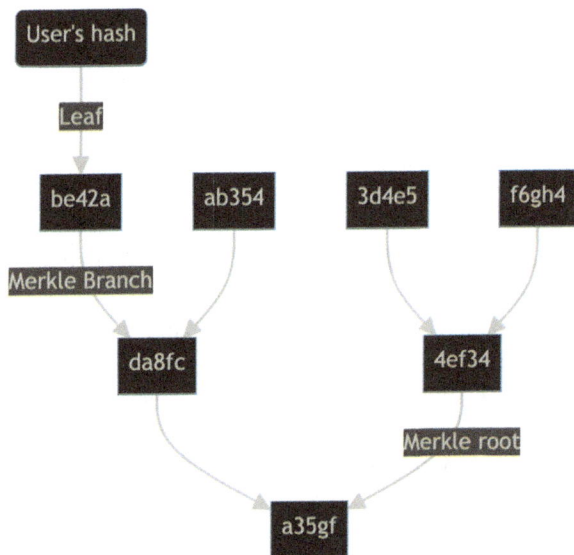

Fig. 4.1 The Merkle Tree

With this system, individuals can cross-check their own balances. They can hash their unique ID with their account balance and locate it within the tree. Several verification rounds ensure the entire tree's integrity, providing assurance for third-party auditors.

4.2.2 Zero-Knowledge Proof

In the Zero-Knowledge Proof system, there's a prover and a verifier. The prover has certain information and wants to convince the verifier of its authenticity without exposing any private details. The prover creates evidence, avoiding sensitive content while underscoring the possession of the said information.

Mathematically, this evidence assures the verifier that the prover genuinely has the information in question. The verifier, upon receiving the evidence, uses a predetermined algorithm to verify its legitimacy without uncovering any confidential details. If verified, the verifier acknowledges the prover's claim without accessing the private details.

As cryptocurrency transactions surge, users seek reassurance about the sufficiency of reserves on exchanges. Yet, the specifics of these reserves and individual assets are typically confidential. Zero-Knowledge Proofs enable exchanges to affirm they're well-resourced and solvent, all without disclosing exact figures. Proof structures, with stringent guidelines, ensure that the derived evidence undergoes thorough computation, eliminating the risk of falsehoods.

By adopting this advanced encryption method, exchanges can boost user trust and increase their financial transparency. This becomes a defense mechanism against potential fiscal vulnerabilities. Unlike traditional third-party audits, zero-knowledge proofs sidestep the need to hand over full reserves to auditors, thus cutting down on operational expenses and potential risks.

4.2.3   The Pitfalls of the Proof of Reserve

After briefly examining the various methods of proof of reserve, it's clear that there are still challenges and shortcomings preventing a foolproof system. Currently, zkSNARKs appear to be the most promising technique for proof of reserves. They ensure user privacy, offer quick verification, and demonstrate the solvency of exchanges. However, confidence in using this method is not yet universally shared.

> For instance, even with Binance's advertised Proof of Reserve, vulnerabilities surfaced. In February 2023, Hacken researchers identified a flaw in Binance's Zero-Knowledge Proof of Reserve (Barwikowski, 2023). Another illustrative case is ZCash, which, in early 2018, identified and subsequently rectified a vulnerability in the type of Zero-Knowledge Proofs they implemented. Surprisingly, this error had existed in the foundational paper guiding the proof's construction for years without detection (Swihart et al., 2019).

While Zero-Knowledge Proofs are technologically advanced and appear promising, concerns are still present about their robustness and reliability, especially under strenuous conditions. In contrast, third-party audits, despite being more traditional, also have their limitations, such as the challenge of sourcing genuine experts and preventing possible collusion between the auditor and their client. Similarly, Merkle trees share comparable drawbacks to Zero-Knowledge Proofs, also Merkle trees require a large subset of user to check if their balance is correct, otherwise the security assumption is compromised.

In essence, while we've made strides in the realm of proof of reserves, there's still a journey ahead to perfect and universally trust these systems.

4.3   Key Management in Crypto Exchanges

In the realm of cryptocurrency, the intricacies of private key management for businesses cannot be overemphasized. These keys are the vital component of any cryptographic system, holding the power to validate, authenticate, and initiate transactions. However, with great power comes great responsibility. The CryptoCurrency Security Standard (CCSS) Matrix, created by the CryptoConsortium, delineates the various layers of considerations that businesses must navigate (CryptoCurrency Certification Consortium, 2022).

Key generation stands as the foundation of any secure system. At a rudimentary level, the generation of private keys depends on pseudo-random number generators. The unpredictability of the source is the most important factor. If the randomness is weak or even partially predictable, it can become the Achilles heel of the system, exposing keys to prediction and potentially disastrous breaches.

Upon key generation, the next step is wallet creation. A wallet in the cryptocurrency ecosystem is analogous to a vault, housing the keys and, by extension, the digital assets. While simple systems may rely on single-signature wallets, the demands of businesses often necessitate multi-signature configurations. This not only distributes trust but also ensures there's no single point of vulnerability. However, orchestrating such configurations is far from trivial. Trust distribution must be balanced with operational efficiency to ensure that transaction processes don't become tediously time-consuming.

Equally vital is the question of how and where these keys are stored. At a basic level, storing keys on internet-connected devices might seem feasible. However, the connectivity of a device is directly proportional to its exposure to threats. The challenges here are numerous: protection against potential malware, ensuring robust encryption standards, and formulating backup strategies that are both secure and easily retrievable.

Any discussion on security would be incomplete without addressing data sanitization. When systems age or are set to be replaced, simply wiping them clean isn't adequate. Data, especially in the digital age, has a way of leaving ghosts—residual traces that skilled adversaries might recover. Ensuring complete, irrecoverable data destruction becomes a necessity.

Audits and certifications introduce another layer of complexity. While periodic self-audits might suffice for smaller operations, larger, more intricate systems benefit from the scrutiny of third-party audits. These external evaluations, especially when conducted by certified professionals, offer a lens of objectivity and rigor. Yet, they can be resource-intensive, both in terms of time and finances.

In the event of breaches or compromises, businesses must be armed with protocols that are both reactive and proactive. Beyond merely notifying stakeholders, there's a need for rapid response plans, automatic system lockdowns, and, crucially, redundant backups that allow for swift recovery and business continuity.

Lastly, the security landscape is ever-evolving. With threats growing in sophistication, defenses must advance in tandem. Basic firewalls and network monitoring are just the starting point. Advanced Intrusion Detection Systems (IDS), Intrusion Prevention Systems (IPS), and real-time threat analyses necessitate substantial investments in both technology and expertise.

In conclusion, the management of private keys in a business context is a journey filled with complexities. Drawing insights from the CCSS Matrix, it becomes evident that mastering this domain requires a blend of technological innovation, operational excellence, and an unshakable commitment to security. As the digital age progresses, so must the defenses guarding its most precious assets.

4.4   Case Studies

4.4.1   FTX: "FTX Is Fine, Assets Are Fine"

In 2019, FTX, a cryptocurrency exchange platform, made its debut and quickly gained widespread popularity. By 2022, it was estimated to have a staggering valuation of $32 billion. Cryptocurrencies, unique due to their encryption and block-chain technology, became attractive investments and payment tools, with some coins, like Shiba Inu, experiencing a jaw-dropping 45 million percent surge in 2021. Platforms like FTX, founded by Sam Bankman-Fried, allowed users to trade these assets and attracted a significant number of customers and heavyweight investors.

However, in November 2022, a storm hit when what seemed like a minor accounting discrepancy unveiled a significant fraud. The misappropriation of cus-tomer funds to Alameda Research, a Hong Kong-based cryptocurrency company, instead of FTX was made public. FTX's rapid rise was further fueled by its aggres-sive strategy of acquiring rivals and high-stakes marketing involving celebrities and high-profile sports sponsorships.

As Bitcoin peaked at $69,000 earlier in 2021, FTX's allure amplified, drawing almost $2 billion in investments. They further expanded their appeal by introducing their native digital token, FTT, in 2019. This token offered several benefits, includ-ing discounts and NFT rewards, to its holders. Yet, the landscape began to shift in late 2021 and early 2022, as cryptocurrency values started to plummet. While many platforms shuttered, FTX appeared resilient, continually expanding.

The tide turned drastically when CoinDesk published an article in November 2022. The piece highlighted Alameda Research's unsettling dependence on FTX's FTT token. Disturbingly, a leaked FTX balance sheet revealed a deep financial intertwining between the two entities, showcasing an alarming negative $8 billion balance. This financial entanglement was further compounded by revelations that Alameda borrowed heavily from FTX's customer funds. The situation became murkier as it was revealed that FTX, being a private entity, did not undergo standard financial audits.

Binance, a competitor but also an early FTX supporter, tried to step in with a buyout. However, they quickly retreated as the depth of the financial mismanage-ment became clearer. FTX's downfall was swift and brutal. Within 10 days in November 2022, the firm lost billions due to massive withdrawals by alarmed cus-tomers, ultimately leading to its bankruptcy. During this chaos, it came to light that Bankman-Fried had misused FTX funds for personal extravagances, marketing splurges, and political donations.

December 2022 saw Bankman-Fried's dramatic arrest on multiple charges, including money laundering and fraud. His subsequent release on a record-breaking $250 million bond set the stage for extensive legal battles, including a class-action lawsuit by distressed FTX investors.

The shockwaves of the FTX scandal rippled across the crypto world. Bitcoin, the flagship cryptocurrency, saw its value fluctuate significantly. On top of that, the

FTX debacle ignited serious discussions among U.S. regulatory bodies about the pressing need for industry regulations. The vulnerability of crypto exchanges was now clearly evident. This incident, coupled with others like the financial difficulties faced by the crypto-based firm Genesis, has injected a dose of skepticism around the stability of cryptocurrencies.

In an effort to make sense of the catastrophe, a detailed report in April 2023 highlighted FTX's myriad operational and managerial failures. The findings were grim, pointing to widespread security lapses, a lack of experienced financial oversight, and a troubling absence of sound financial protocols. The company's careless approach to intercompany transactions was particularly concerning (Ferreira, 2023; Hetler, 2023).

4.4.1.1 Exchange Platform Token Management: Risks of Over-Reliance

Cryptocurrency exchanges have increasingly been seeking to diversify their offerings and retain users within their ecosystems by launching proprietary platform tokens. These tokens often come with perks such as reduced trading fees, access to premium services, and voting rights on platform developments. While they present an attractive proposition for both exchanges and their users, the case of FTX's FTT token and its collateral use has illuminated potential risks tied to over-reliance on such tokens.

FTX, a renowned cryptocurrency exchange, introduced its native token, FTT. Similar to other exchange tokens, FTT offered multiple benefits for its holders, including fee reductions and increased referral bonuses. However, a unique feature that distinguished FTT was its function as collateral for future positions on the FTX platform. This meant traders could leverage their FTT holdings to open positions in other cryptocurrencies or financial instruments, increasing the utility and, theoretically, the demand for FTT.

On paper, this seems like a logical evolution, combining the utility of an exchange token with the capabilities of collateral assets. However, it brings to the forefront a complex interplay of factors that could jeopardize both the stability of the token and the broader platform. The intricate relationship between FTT's price, its use as collateral, and the overall health of the FTX exchange created a potential feedback loop of vulnerabilities.

The risk became evident when traders started heavily relying on FTT as collateral for significant loan positions. In a stable or bullish market, the model appeared resilient. But the crypto market, known for its volatility, doesn't always favor the bullish narrative. In a market downturn, where the value of FTT might decline, those leveraging FTT as collateral could face margin calls, necessitating additional collateral or forcing the liquidation of their positions.

Such forced liquidations, especially if they occur en masse, could further depress the price of FTT. This would lead to a cascading effect, where the falling value of the token triggers more margin calls and liquidations. The feedback loop becomes a vicious cycle, amplifying the negative price action and creating a potential "death spiral." Moreover, with FTT's price in turmoil, it could erode trust in FTX's ecosystem, leading users to question the stability and viability of other services tied to the token.

The broader implications of such a scenario are worth noting. First, the exchange itself might face liquidity challenges, especially if it has to step in to prevent cascading liquidations. The situation can become particularly dire if large-scale liquidations coincide with a broader market downturn, straining the exchange's resources. Second, it might cause reputational damage, not just to FTX but to other exchanges that may employ similar models. Users, fearing similar vulnerabilities, might become hesitant to engage with exchange tokens, even on platforms unrelated to FTX.

This situation underscores the importance of careful risk management and the dangers of over-relying on a single token or financial instrument. While diversification and innovation are commendable, they must be approached with a thorough understanding of potential systemic risks.

In conclusion, the FTT scenario serves as a warning about the complexities of integrating platform tokens into broader financial structures. As the crypto space matures, exchanges and platforms must continuously evaluate the risk-reward equation, ensuring that in the pursuit of innovation, they don't inadvertently introduce vulnerabilities that could undermine the very ecosystems they seek to improve.

4.4.2   Mt. Gox: The Downfall of an Icon

Founded in Tokyo, Mt. Gox dominated the cryptocurrency world between 2010 and 2014, handling over 70% of global Bitcoin transactions at its peak. Created by Jed McCaleb, it originally catered to "Magic: The Gathering" card game enthusiasts. Its name, in fact, stands for "Magic: The Gathering Online Exchange." Mark Karpeles took over in 2011, becoming the principal shareholder and CEO.

Under Karpeles, Mt. Gox reached staggering heights, becoming the most significant Bitcoin exchange worldwide. This dominance granted it an unparalleled influence on Bitcoin's market activity. Case in point: in 2013, when it briefly halted trading to stabilize the volatile market.

However, its prominent status made Mt. Gox an attractive target for cyberattacks. The exchange suffered security breaches several times, with hackers using various tactics to exploit vulnerabilities. In 2011, they capitalized on stolen credentials to illicitly transfer Bitcoins, while later that year, network protocol weaknesses led to the loss of thousands more.

Trouble escalated in early 2014. Technical glitches plagued the system, making it impossible for Mt. Gox to determine transaction details with confidence. These problems culminated in February 2014 when it suspended withdrawals due to "suspicious activity." The discovery of the loss of a colossal number of Bitcoins, somewhere between 650,000 and 850,000, soon followed. While they managed to recover

200,000 Bitcoins, the remaining loss was crippling. The monetary worth of the lost Bitcoins reached into the hundreds of millions, forcing Mt. Gox into insolvency. It filed for bankruptcy shortly afterward.

Post its collapse, the remnants of Mt. Gox became mired in legal complications. The estate, comprising more than 200,000 Bitcoins and Bitcoin Cash, was under the supervision of trustee Nobuaki Kobayashi. Rumors abounded regarding the involvement of Russian hackers in the Bitcoin heist, while CoinLab, a significant creditor, pursued a breach of contract lawsuit worth billions against Mt. Gox. Mark Karpeles, its CEO, faced legal consequences too, found guilty in 2019 of falsifying data.

Closure seemed to come in November 2021. Mr. Kobayashi announced an agreement between Japanese courts and Mt. Gox creditors concerning the rehabilitation plan, marking the end of a long-standing legal battle. This plan outlined a phased compensation strategy for various creditors. The system allowed only those with approved rehabilitation codes to register, with no provisions for new claims.

The fate of Mt. Gox in the cryptocurrency landscape remains uncertain. Its once-lofty presence has been reduced to a reminder about the risks inherent in the digital currency domain. Only a fraction of the lost Bitcoins were recovered, with the majority either lost forever or removed from circulation. Today, Mt. Gox exists in limbo. Its website has been inactive since 2014, and its future in the crypto industry is still up in the air (Frankenfield, 2023).

4.4.3   QuadrigaCX: Mismanagement and Disappearance

In the unpredictable world of cryptocurrency, the story of QuadrigaCX stands out. Once lauded as Canada's premier crypto exchange, its dramatic fall began with the sudden death of CEO Gerald Cotten, subsequently leading to a loss of approximately $190 million in user funds.

Gerald Cotten wasn't merely the CEO; he was the lifeblood of QuadrigaCX's operations. His unexpected demise in India due to Crohn's disease complications catapulted the exchange into chaos. Most notably, Cotten was the only one given access to the exchange's cold wallet keys—essentially, the encrypted passwords to a treasure of $145 million in cryptocurrencies. This exclusive access became the epicenter of the crisis, as without Cotten, the funds seemed irretrievable.

Cotten's widow, Jennifer Robertson, didn't escape the spotlight either. Following her husband's death, their combined property acquisitions, totaling 16 diverse properties, became a point of contention. To further complicate matters, Robertson expedited the restructuring of QuadrigaCX's leadership, a move that only deepened suspicions.

Interestingly, for an exchange of its stature, QuadrigaCX had an anomaly: its one and only developer, Alex Hanin. While typical crypto exchanges employ a plethora of developers to ensure smooth operations, QuadrigaCX leaned heavily on Hanin. Such a decision was not only risky but also signaled potential vulnerabilities and shortcuts in the system.

The puzzle also brims with pressing questions. The authenticity of Cotten's death became a matter of debate. Though a doctor confirmed his demise from Crohn's complications, the circumstances were too opportune for doubters. The exchange was in a financial abyss, and Cotten's death seemed almost too timely, leading some to even demand the exhumation of his body for confirmation.

The whereabouts of the $145 million in cryptocurrency added to the enigma. Initially believed to be locked in cold storage, subsequent audits painted a different picture: the wallets were nearly empty. Funds had been maneuvered to other accounts, possibly other exchanges, obscuring their trail and complicating their recovery.

Several theories have emerged in the wake of these mysteries. Some hypothesized that QuadrigaCX was already sinking, with a 2017 bug locking a vast amount of its Ethereum. Cotten's death might have been a smokescreen to mask these pre-existing issues. Others, pointing to Patryn's dubious past, theorize a money laundering operation, especially given the exchange's extensive cash withdrawals. Then there's the possibility of sheer incompetence, suggesting that poor management, a lack of oversight, and inadequate staffing doomed the exchange.

Regardless of which theory holds water, the QuadrigaCX incident underscores the dangers waiting in the untested waters of cryptocurrency. With hundreds of crypto exchanges operating globally, many without stringent regulations, the stage is set for potential repeats of such calamities. As the QuadrigaCX investigation unfolds, it serves as a warning to investors and a call for improved oversight in the crypto world (Copeland, 2020).

4.4.4 Bitfinex: Heist of the Century

The 2016 Bitfinex hack is often called the "heist of the century." This Hong Kong-based cryptocurrency exchange suffered a major attack in August 2016, leading to the theft of 119,754 bitcoin. At the time, this amounted to a loss of $72 million, but with bitcoin's increased value, the stolen sum has surged to a staggering $8.2 billion.

The hack's sophistication was evident when the attackers bypassed the multisig wallet security protocols that Bitfinex employed with their custodian partner, BitGo. Many questioned Bitfinex's decision to not use cold storage, leading to speculations about potential inside involvement. In response to this crisis, Bitfinex's strategy was twofold: distribute the loss across all accounts, resulting in about a 36% reduction in assets for every user, and compensate affected users with BFX tokens, equivalent to their losses. These tokens could later be exchanged for shares in Bitfinex's parent company, iFinex, or redeemed once funds were recovered.

This decision was met with mixed reactions. While some appreciated Bitfinex's approach, others were unhappy, especially those who hadn't lost any assets in the hack. Despite the setbacks, Bitfinex's commitment led to the full redemption or trading of all issued BFX tokens within a year.

The aftermath of the hack saw Bitfinex working alongside global law enforcement agencies to recover the stolen bitcoins and identify the culprits. The journey to justice had its twists: a minor recovery of 27 bitcoins in 2018, the arrest of two Israeli brothers, Eli and Assaf Gigi, in 2019 on suspicion of their involvement, and the movement of a significant portion of the stolen bitcoins in 2020.

Fast forward to February 2021, and the U.S. Department of Justice (DOJ) announced the largest-ever confiscation of stolen cryptocurrencies, retrieving 94,636 bitcoins worth around $4.5 billion. This action followed extensive investigations by both the IRS Criminal Investigation department and the FBI. A significant development was the arrest of a New York couple, Heather Morgan and Ilya "Dutch" Lichtenstein. They are suspected of attempting to launder the proceeds from the hack. Interestingly, the bulk of the stolen bitcoins remained untouched in a digital wallet for approximately 6 years. Only when attempts were made to transfer these funds into the mainstream banking system did authorities successfully trace and link the transactions to real-world identities.

The couple now faces serious charges. Both are accused of money laundering, an offense carrying a maximum penalty of 20 years in prison. Additionally, they are charged with conspiracy to defraud the USA, which could lead to another 5 years in jail. Officials allege that the stolen bitcoins were sent to Lichtenstein's digital wallet. The tracing of these transactions underscores the capability of specialized software and blockchain analytics providers, which can exploit the blockchain's transparent nature. Every transaction is publicly recorded, making it possible, with the right tools, to track even illicit transfers (CoinBureau, 2023).

4.4.5   Coincheck: The Perils of Hot Wallet Reliance

This breach saw an astounding $523 million in NEM coins (XEM) being redirected to an unknown address from Japan's Coincheck cryptocurrency exchange on January 26, 2018. Despite intensive investigative efforts, the identities of these anonymous hackers, as well as the lost assets, remained concealed in mystery.

The financial ramifications of this intrusion were staggering, amounting to approximately $500 million in losses. Coincheck, initially confident about recovering these assets, later confessed to an unsettling revelation: their defenses were compromised owing to staffing shortages at that time. This lack of manpower inadvertently granted these hackers unrestricted access to their systems. However, the saga did not end in despair. Several hours after the intrusion was detected, Coincheck, in a move of the highest integrity, leveraged its own financial reserves to compensate the 260,000 customers affected by this hack. This act of restitution, combined with the introduction of enhanced security protocols, played a crucial role in rehabilitating Coincheck's damaged image (CoinMarketCap, 2022).

4.4.6   Binance: Tackling a Sophisticated Attack with Proactive Measures

In a significant security breach, cryptocurrency exchange Binance reported a loss of 7000 BTC, equivalent to about $40.7 million. This malicious act, described as a "large-scale security breach," was meticulously executed by hackers who managed to access sensitive information such as user API keys and two-factor authentication codes. Former Binance's CEO, Changpeng Zhao, communicated this unfortunate incident, emphasizing the sophisticated methods employed by the hackers. The illicit withdrawal went unnoticed, successfully bypassing existing security checks, which raised alarms only after the transaction was completed.

The breach's revelation was strategically managed by Binance, with Zhao initially announcing some "unscheduled server maintenance," assuring users that their funds were secure. Detailed communication followed this preliminary message, explaining the breach's nature and impact. Despite the substantial loss, the breach seemed to have affected only Binance's hot wallet, accounting for about 2% of the exchange's total bitcoin holdings. The remaining wallets were untouched, remaining secure from the hackers' infiltration.

Zhao's transparent communication revealed the complex approach of the hackers, who displayed strategic patience, executing their actions with precision through multiple accounts that seemed unrelated. The timing and the transaction structure were such that it managed to bypass the sophisticated security checks that were in place. Subsequent to the identification of this breach, Binance took immediate corrective action. Withdrawals were temporarily frozen to assess the breach's impact and contain further possible losses. Zhao further indicated that while the affected accounts were being scrutinized, there might still be some accounts under the hackers' control.

In the breach's aftermath, Binance undertook a commitment to a comprehensive security review. This review would entail a meticulous examination of the systems and data to enhance the exchange's security architecture. To reassure and protect the users from the effects of this breach, Binance activated its Secure Asset Fund for Users (SAFU fund). This fund, built by allocating 10% of all trading fees, acts as a protective shield in unprecedented situations, ensuring that users aren't adversely affected by such losses.

The former CEO, Zhao, in a display of transparency and commitment to user trust, maintained active communication. He welcomed user queries and engagement, even participating in a scheduled Twitter "ask-me-anything" session. This incident illustrates the dynamic challenges that the cryptocurrency exchanges deal with, where, despite robust security architectures, malicious actors sometimes manage to execute significant breaches. It also underscores the resilience and strategic crisis management that platforms like Binance employ to protect and reassure their user base in the face of such adversities (CoinDesk, 2021).

4.4.7 Coinbase: Tree of Alpha Saves the Day

In a swift act of ethical vigilance, "Tree of Alpha," a reputed White Hat hacker, reached out to Coinbase's CEO, Brian Armstrong, cautioning him of a potential vulnerability. This flaw, if exploited, could have allowed malicious entities to manipulate Coinbase order books, leading to significant financial gain for nefarious actors. Within a short span after this revelation, Coinbase issued a statement, confirming they had paused trading on their Advanced Trading platform due to this technical hiccup.

> Sincere gratitude is extended to Tree of Alpha and fellow whitehat hackers. Their efforts, often under-recognized, play a crucial role in fortifying web3's security landscape. Tree of Alpha, among others, deserves acknowledgment for making web3 a safer domain. Sincerely, a note of appreciation.

Once the issue was addressed and rectified, both Tree of Alpha and Armstrong exchanged words of appreciation. While the hacker lauded Coinbase's speedy corrective measures, Armstrong acknowledged Tree of Alpha's invaluable assistance during this crucial point in time. The proactive stance of Coinbase became evident when they acted promptly on Tree of Alpha's alert, suspending trading within 2 h of the tweet's disclosure. Mere hours later, normal services, including advanced retail features, resumed. Tree of Alpha, for added assurance, corroborated the fix with an illustrative screenshot showcasing the exploit's mitigation. Armstrong's subsequent shout-out to Tree of Alpha emphasized the sense of community that the crypto domain occasionally displays (Encila, 2022).

4.4.8 KuCoin: Navigating a Major Breach and Demonstrating Resilience

In September 2020, the cryptocurrency world was rocked by one of the most significant breaches of a trading platform. KuCoin, a widely-recognized cryptocurrency exchange based in Singapore, experienced a devastating security breach. This resulted in unauthorized withdrawals amounting to a staggering $281 million, highlighting the vulnerabilities inherent in even major exchange platforms.

The illicit operation was brought to light when KuCoin's surveillance systems detected irregular and unauthorized large withdrawals of bitcoin (BTC) and ethereum (ETH) tokens. These withdrawals were being funneled into an unrecognizable and suspicious wallet. As the gravity of the situation started becoming evident,

Johnny Lyu, KuCoin's CEO, took to a live stream to communicate with the exchange's user base and the wider cryptocurrency community. Lyu candidly revealed that cyber attackers had somehow obtained the private keys to KuCoin's hot wallets, the online storage systems for the platform's assets.

In response to the emerging crisis, KuCoin rapidly implemented damage control protocols. The affected assets were immediately shifted from the compromised hot wallets to new secure ones. The exchange also took the precautionary step of temporarily freezing user deposits and withdrawals to prevent additional unauthorized transfers. One silver lining in this tumultuous situation was that KuCoin's cold wallets, offline storage systems widely regarded for their robust security measures, remained untouched and secure, ensuring that a significant portion of the platform's assets remained safe.

The aftermath of the hack had broader repercussions in the cryptocurrency ecosystem. Recognizing the severity of the breach, other cryptocurrency exchanges and asset managers, notably Bitfinex, acted in solidarity with KuCoin. They promptly blacklisted the suspicious wallet addresses associated with the hack, rendering the stolen assets virtually unusable in the mainstream cryptocurrency market (Hui & Zhao, 2021).

4.4.9 Houbi: A Costly Mistake

HTX, a notable cryptocurrency exchange previously known as Huobi, has recently faced a security breach resulting in the loss of 500 Ether (ETH), an amount valued around $8 million. The disclosure was made by Justin Sun, an advisor at HTX and the founder of Tron. He clarified that the unfortunate event took place on a Sunday and that the identification of the breach was immediate, allowing for a quick response.

According to Sun's statement on the social media platform X, formerly known as Twitter, the compromised wallet was one of the hot wallets of HTX. Historical data shows substantial activity in this wallet, having received an approximate $500 million in deposits from Binance since its inception in March, as per the details provided by Arkham data.

The hack, while significant, was downplayed by Sun, who offered some reassurances and perspectives on the extent of the loss. He mentioned that the $8 million loss, while substantial, represents a relatively minor portion when compared to the $3 billion worth of assets that the exchange holds for its users. Putting it further into perspective, Sun explained that the loss equated to about 2 weeks' worth of revenue for HTX. Consequently, this allowed the company to cover the losses fully, ensuring that the funds and the trading operations remained secure and continued seamlessly. Sun emphasized that all necessary actions were taken promptly, ensuring that the platform's regular operations and user funds remained unaffected and secure.

In a rather unconventional move, Sun revealed that HTX is not only willing to offer the hacker a $400,000 bug bounty for returning the stolen funds, but also expressed openness to hiring the hacker as a security advisor, specializing in white hat operations. This approach indicates a willingness to convert a crisis into an opportunity to strengthen the exchange's security framework, turning a potential adversary into an ally to improve the platform's resilience against future security threats.

4.4.10 Summary of Case Studies

The chronicles of Mt. Gox, Bitfinex, KuCoin, HTX, Coinbase, Binance, Coincheck, Quadriga, and especially FTX, yield profound insights that inform both the builders of these platforms and their myriad users.

The saga of Mt. Gox remains a lesson on the perils of neglect and complacency. Its catastrophic collapse isn't just a singular event but a resonating lesson on the importance of vigilance in an industry where stakes are sky-high. On the other hand, KuCoin's handling of its breach showcases the power of swift action, transparency, and adaptability. The episode illustrates that while technological fortifications are critical, an ethical stance, combined with effective communication, holds equal merit.

Bitfinex, in its recovery from adversity, epitomizes the spirit of resilience. Their story underscores that setbacks, while painful, can be transformed into opportunities for growth, innovation, and strengthening of partnerships. HTX, with its unorthodox approach to its security breach, brings forth an interesting perspective on how crises can sometimes be turned on their heads.

The challenges faced by titans like Coinbase and Binance serve as reminders that no entity, regardless of its stature, is immune from pitfalls. Their stories highlight the pressing need for continuous user education, emphasizing that while platforms must be steadfast in security, users must be equally informed and vigilant.

The situation at Coincheck stands as an evident reminder of the dangers of excessive reliance on hot wallets. Their episode drives home the point about the balance between convenience and security. In the meantime, the enigmatic and unsettling story of Quadriga highlights how technological vulnerabilities and human factors are inextricably linked. It stresses the importance of comprehensive operational foresight.

However, among all, the narrative of FTX stands out as the most haunting. Contrary to its peers, FTX's journey wasn't just marred by technological vulnerabilities or operational oversights but was steeped in profound tragedy and loss. It serves as a grim testament to the catastrophic consequences that can occur when oversight, complacency, and lack of due diligence converge. The tragic lesson that can be learned from FTX's story is that mistakes can have irreversible human as well as financial consequences.

In essence, the world of cryptocurrency exchanges, with its tumultuous fluctuations, calls for a comprehensive approach to security, ethics, and user engagement. As we navigate this complex landscape, may the tales of these exchanges serve as beacons, highlighting both the pitfalls to avoid and the standards to aspire for.

4.5 Conclusion

This chapter has explored various crucial facets pertaining to the security of centralized crypto exchanges. We discussed proof-of-reserve techniques like Merkle trees and zero-knowledge proofs that enable exchanges to demonstrate solvency without compromising sensitive data. Additionally, we delved into best practices for private key management, underscoring the importance of robust protocols across the key lifecycle.

However, these technological safeguards form only one piece of the puzzle. The case studies analyzed reveal that oversight, complacency, and lack of transparency can override even the most sophisticated security architecture. The spectacular downfall of giants like Mt.Gox and FTX serve as stark reminders that technological resilience must be matched by ethical operations, financial transparency, and comprehensive risk management.

As the crypto industry continues to evolve amidst volatility and adversity, the path ahead necessitates a multidimensional approach. Exchanges must champion innovation in privacy-preserving proofs of reserves while simultaneously investing in experienced leadership to promote operational excellence. Equally vital are clear regulatory guidelines to enable better oversight without stifling growth.

In essence, securing centralized crypto exchanges in the digital era requires ceaseless vigilance, technological competence, financial transparency, and a steadfast commitment to ethical values. It is only with a holistic vision encompassing these crucial elements that the true potential of Web3 can be sustainably harnessed for the world. Therein lies the key to constructively balancing the immense promises and complex challenges ushered in by this financial revolution.

References

Barwikowski, O. M. (2023, September 14). Binance's Proof of Reserves gets a security boost thanks to Hacken's discovery - Hacken. Hacken. https://hacken.io/case-studies/binance-discovery/
CryptoCurrency Certification Consortium (2022, June 28). CCSS Table - CryptoCurrency Certification Consortium (C4). https://cryptoconsortium.org/ccss-table/
CoinMarketCap. (2022). Coincheck Hack - One of The Biggest Crypto Hacks in History. CoinMarketCap Academy. https://coinmarketcap.com/academy/article/coincheck-hack-one-of-the-biggest-crypto-hacks-in-history.
Copeland, T. (2020, February 12). The complete story of the QuadrigaCX $190 million scandal. Decrypt. https://decrypt.co/5853/complete-story-quadrigacx-190-million

CoinDesk. (2021, September 13). Hackers steal $40.7 million in bitcoin from crypto exchange Binance. CoinDesk. https://www.coindesk.com/markets/2019/05/07/hackers-steal-407-million-in-bitcoin-from-crypto-exchange-binance/

CoinBureau. (2023). Bitfinex Hack: What Happened, Who Did it and What's the Latest? Coin Bureau. https://www.coinbureau.com/analysis/bitfinex-hack/

Encila, C. (2022, February 13). White Hat Hacker Protects Coinbase from Danger by Spotting Threat | Bitcoinist.com. Bitcoinist.com. https://bitcoinist.com/white-hat-hacker-saves-coinbase/

Ferreira, P. (2023, September 5). The FTX full story: All you need to know. Financial and Business News | Finance Magnates. https://www.financemagnates.com/cryptocurrency/the-ftx-full-story-all-you-need-to-know/

Frankenfield, J. (2023). What was Mt. Gox? Definition, history, collapse, and future. Investopedia. https://www.investopedia.com/terms/m/mt-gox.asp

Hetler, A. (2023). FTX scam explained: Everything you need to know. WhatIs.com. https://www.techtarget.com/whatis/feature/FTX-scam-explained-Everything-you-need-to-know#:~:text=FTX%20crashed%20due%20to%20mismanagement,of%20mishandled%20customer%20funds%20surfaced

Hui, A., & Zhao, W. (2021, September 14). Over $280M Drained in KuCoin Crypto Exchange Hack. Coindesk. https://www.coindesk.com/markets/2020/09/26/over-280m-drained-in-kucoin-crypto-exchange-hack/

Swihart, J., Winston, B., & Bowe, S. (2019). ZCash counterfeiting vulnerability successfully remediated. Electric Coin Company. https://electriccoin.co/blog/zcash-counterfeiting-vulnerability-successfully-remediated/.

Carlo Parisi is a senior Solidity smart contract developer, senior auditor, and content creator with a degree in Computer Science. He has a deep knowledge in developing and auditing Solidity code, the main language used in Ethereum for smart contracts. He has been a Bitcoin Enthusiast since 2013, DeFi user since 2018.

Dmitriy Budorin is Founder & CEO at Hacken and Founder at HackenProof.
Dyma is a cybersecurity expert and crypto economy influencer with 14+ years of managerial expertise in cybersecurity as well as risks and controls audits. Dyma holds a master's degree in International Economics and an MBA from the Kyiv Institute of Investment Management. He is a certified member of the Association of Chartered Certified Accountants (ACCA).
In 2017 Dyma established Hacken, a cybersecurity consulting firm. Five years later, Hacken is trusted by the largest crypto projects; the company's portfolio includes HackenAI, HackenProof, CER, and a suite of accompanying blockchain services. Dyma's effective leadership is what transformed Hacken from a startup into a major player in Web3 cybersecurity. The story of success is only gaining momentum.
As the company's Co-Founder and CEO, Dyma is responsible for leading the team of 100+ talented specialists and providing a vision of the future. Dyma consults the Ukrainian government on the adoption of a virtual economy. He is a regular participant in major Web3 cybersecurity events as an invited speaker.

Chapter 5
CBDC Security

Zhijun William Zhang

Abstract Central bank digital currencies (CBDCs) are emerging as a major player in digital payments as well as digital finance in general. As a new financial instrument offered by a central bank, a CBDC can be quite complex and require the central bank to collaborate with a large number of market participants, thus introducing a large attack surface. To secure its CBDC, a central bank not only needs to secure the core CBDC design and its supporting infrastructure managed by the central bank but also take a leadership role in ensuring a secure ecosystem. In addition, the use of novel technologies such as blockchain and smart contract could introduce security threats that are new to the central bank and other participants, which would require special attention. In addition, CBDCs bring about new ways for financial crimes and could pose a threat to data privacy of CBDC users. These risks must be recognized during the CBDC design, with countermeasures established along the lines of people, process, and technology.

For the Web 3 ecosystem to function, digital payments are essential. Compared to private money such as digital currencies, stable coins, and central bank digital currencies (CBDCs) are public goods providing such utilities. As a digital form of fiat currency, a CBDC is a direct liability to the central bank, with its value backed by the central bank. It is designed to be accepted by all—anyone who has received CBDC as payment can in turn either spend it or convert it to a bank account balance through its bank or other CBDC intermediaries. Most central banks make a distinction between wholesale CBDCs that will only be made available to financial institutions and retail CBDCs that are meant for all consumers, businesses, and merchants. From a security perspective, retail CBDCs have a much wider attack surface than wholesale CBDCs. This book chapter will focus on retail CBDCs, as its security measures tend to be a superset of what is required to secure a wholesale CBDC.

Views presented in this book chapter are the author's own, and do not represent the official position of the Bank for International Settlements (BIS) or the BIS Innovation Hub.

Z. W. Zhang (✉)
Bank for International Settlements Innovation Hub, Stockholm, Sweden

© The Author(s), under exclusive license to Springer Nature Switzerland AG 2024
K. Huang et al. (eds.), *Web3 Applications Security and New Security Landscape*, Future of Business and Finance, https://doi.org/10.1007/978-3-031-58002-4_5

Specifically, this chapter will cover the following topics:

Risks and threats facing CBDCs: Analyzing the risks and challenges of CBDCs, including regulatory, technical, and cyber security risks, and how they can be mitigated.

The importance of secure infrastructure in CBDC development: Discussing the need for robust security measures in the development and deployment of CBDCs, including measures such as encryption, multi-factor authentication, and secure storage solutions. This is further divided into security of the core CBDC design, and security of the ecosystem that supports CBDC operations.

CBDC and the role of central banks: Examining the role of central banks in ensuring the security of CBDCs, including their role in developing and implementing security protocols, as well as their potential use as a backstop in the event of a security breach.

CBDC and blockchain technology: Exploring the potential use of blockchain technology in CBDCs and its implications for security, including the benefits of decentralization and immutability, as well as the potential challenges and risks.

CBDC and financial crime: Discussing the potential impact of CBDCs on financial crime and the measures that can be taken to prevent it, such as anti-money laundering (AML) and know-your-customer (KYC) regulations.

CBDC and privacy: Examining the privacy implications of CBDCs, discussing how they might be used to support financial inclusion and protect individuals' personal data.

5.1   Risks and Threats Facing CBDCs

The Internet and the digital world have been exploited by malicious actors of different kinds for various purposes, for which financial gains have been a major motivation. As an example, the estimated total loss in 2023 across cryptocurrency networks was over 1.8 billion US dollars (CertiK, 2024). This represents a significant decrease from 2022's total of over 3.7 billion in 2022, when there were much more cryptocurrency activities.

CBDCs, with its vast user base and a variety of stakeholder involved, tend to have a large attack surface. Attacks can range from regular cyber criminals compromising a wallet or payment process for financial gains, sophisticated attackers compromising the CBDC design to repeatedly double-spend or counterfeit CBDCs, to a malicious nation-state actor trying to use the weakness in a nation's CBDC system to cause major disruption in payment systems or compromise financial stability by manipulating CBDC supply, etc.

Central banks must continuously analyze the potential threat actors, the threats they pose, and whether the domestic CBDC system is vulnerable to any of such threats. Below is a non-exhaustive list of threats that CBDC systems may face (The Atlantic Council, 2022; BIS Innovation Hub, 2023a, 2023b, 2023c, 2023d, 2023e,

2023f; BIS Representative Office for the Americas, 2023; The Digital Dollar Project, 2023):

1. Distributed denial-of-service (DDoS) attacks: An attacker can use computing resources they control, including IoT devices, to launch a massive volume of service requests to a CBDC system's core component or a service it relies on, in order to exhaust the computing resources in a critical area of the ecosystem, resulting in a system overload that causes failure, timeouts, or significant performance degradation.

2. Advanced persistent threat (APT) attacks: APT attacks are typically launched by nation-states or organized crime groups. They penetrate the victim's system, plant malicious software or create backdoors, and patiently observe network traffic and user and system behavior, sometimes lying dormant for a long period of time. APTs typically employ advanced techniques to evade intrusion detection technologies, and quietly exfiltrate data from the victim's network or move laterally to gain control of additional computing resources. Eventually, they could steal significant amounts of money through fraudulent transactions, or compromise a large amount of sensitive information.

3. Counterfeiting or double spending of CBDCs. An end user of a CBDC system or a criminal group could try to defraud CBDC payment devices or mobile applications to devise ways to double-spend CBDCs in their possession, or to counterfeit CBDCs. Offline wallets will likely give attackers extra time to try different techniques while not being detected while the device remains not connected to the network.

4. Malware (including ransomware) attacks: An attacker plants malicious software into the target's computers and networks, which subsequently could either destroy certain computing services, become a backdoor for attackers to connect to the victim's network, or be used to hold the victim's information and computing assets hostage for ransom payments. These can remain dormant and hidden until required by an attacker.

5. Social engineering attacks: An attacker could use techniques such as phishing, spear-phishing, SIM swaps, man-in-the-middle, or compromised credentials to take control of an end user's CBDC account or a privileged user's account that is used to manage a part of the CBDC system.

6. Cryptographic key compromise: A malicious actor could obtain a user or organization's private key for claiming ownership of CBDCs by hacking the computer or device containing the key file, searching through the device's memory for traces of the key, conducting cryptanalysis based on collected data that have been generated using the key, or via special side-channel attacks, etc.

7. Attacks against new technology components related to DLT or smart contracts: An attacker could find and exploit vulnerabilities in smart contracts, in a DLT consensus protocol, in cross-ledger bridges, in oracles or in governance protocols, etc. that are inherent for a DLT solution but may not be familiar to central banks or commercial banks who operate different part of the CBDC ecosystem.

8. Compromise of the payment process: As a payment process often involves multiple parties, the logic in the series of steps involved could have logical gaps, or lack security safeguards, which could be exploited by an attacker to make a purchase without paying, redirect payments to a different recipient, replay payment instructions, harvest payments from wallets that do not require payer consent, or trigger a refund without first making the payment.

9. Insider attacks: These include sabotage or fraud. Insider sabotage could happen when a disgruntled employee or contractor who has access to a CBDC system attempts to cause the system to malfunction by damaging the hardware, deleting key information, shutting down services, providing incorrect input, or enabling other threat actors. Insider fraud could occur when a malicious insider acts individually or together with other threat actors to commit financial fraud. The attacker could leverage their privileged access and knowledge of the CBDC system's business logic, and devise ways to defraud the system.

10. Human error, negligence, or lack of awareness: This threat could materialize in different forms. For example, a developer may include an open-source package with security vulnerabilities in the application; the operations team may have delayed applying a security patch or other critical updates; a system administrator may mistype a command during system maintenance or forget to renew an expiring digital certificate.

11. Information disclosure due to lack of proper controls: Data related to users and transactions may be stored across different financial intermediaries, with some (hopefully aggregated or anonymized) data stored at the central bank. An employee at such institutions may gain unauthorized access due to poor or lack of access management controls. A third-party service provider for these institutions may see sensitive information related to the CBDC system due to a misconfiguration of the shared IT environment it may have access to, or during troubleshooting when regular controls are not effective. Data could also be exposed to attackers due to control gaps in these organizations' networks.

12. Attack against supply chain vendors or service sub-vendors: Service and solution providers involved in the supply chains for components of a CBDC system, such as software or hardware, cloud or data center service providers, or as part of the overall CBDC life cycle such as user onboarding agent, could be targets of threat actors. Any compromise could lead to impacts on the integrity or availability of CBDC systems, confidentiality of sensitive data in the systems, or integrity of the user identity screening process.

13. AI-assisted attacks: With the fast advancement of AI technologies, threat actors are leveraging such technologies or any weakness in them to devise new types of attacks. Such attacks could result in compromise and even takeover of CBDC wallets or privileged accounts for managing the CBDC system, leakage of sensitive user and payment data, or loss of integrity of payment transactions, etc. Central banks must monitor the CBDC system closely in order to detect such novel attacks as early as possible.

Risks from such threats include financial, operational, and reputational. Most threats are out of the central bank's control. However, central banks can assess whether their CBDC systems are vulnerable to each threat, and if so, what the likelihood and impact this threat could cause. Central banks need to minimize their CBDC systems' vulnerabilities to the point where either the likelihood of a compromise happening and its impact are deemed small enough to be acceptable, or the central bank has countermeasures to mitigate the risk. Such analysis should be repeated regularly (no less than once a year) to account for any changes in the threat landscape or the CBDC system, including the availability of new technologies to make certain compromises more likely, or new vulnerabilities as a result of an expansion of the attack surface.

5.2   Core CBDC Design and Security

As central bank digital currencies (CBDCs) represent a new form of central bank money, it is critical to establish a secure supporting infrastructure before launch. This infrastructure underpins the trust in the system for all involved entities including the central bank, commercial banks, payment providers, and end users. A robust security posture needs to be embedded across the ecosystem's components like the network, data stores, wallets, and more. Additionally, effective controls have to be instituted to secure CBDC operations and transactions. This section outlines recommendations on infrastructure security, supply chain risk management, securing participant systems, and core CBDC architectural decisions to prevent fraud or attacks that could undermine confidence in a CBDC.

5.2.1   Establishing a Secure Infrastructure

Since the CBDC ecosystem typically involves the central bank, commercial banks, payment service providers, and other service providers (Auer & Böhme, 2020, #), it is important to establish a trust relationship among these entities. Establishing a permission blockchain network is one way to achieve this. The traditional public key infrastructure (PKI) approach could also be leveraged for the different players to trust each other in performing CBDC operations (Syed & Lee, 2022).

In this PKI, the central bank would typically manage the root keys, and use its private key to sign digital certificates for the other entities. Intermediaries such as commercial banks would use their private keys to generate digital certificates for CBDC wallets, which in turn should store the public keys from both the central bank and approved intermediaries in their key store. During daily use, CBDC wallets can use these public keys to validate the integrity of a digital certificate presented by a counterparty CBDC wallet, similar to how a browser would validate the digital certificate presented by a web server in a TLS protocol (Cockroach Labs, 2020).

Similar to other solutions, central banks need to secure the network, servers, and storage technology components that support the CBDC solution, applying advanced security measures from firewalls, intrusion detection, data encryption in transit and at rest, and around-the-clock monitoring and alerting. It is also very important to enforce security best practices for privileged users, including least privilege, just-in-time authorization, and multi-factor authentication. In addition, the central bank should look for opportunities to automate its security controls including the adoption of DevSecOps with infrastructure-as-code (US Department of Defense, 2021) and the establishment of security guardrails.

Supply chain security is an area that central banks need to pay attention to. This ranges from the computing infrastructure providers, data feed providers, and software modules and libraries that the CBDC system depends on. A robust third-party management program must be established, with software modules validated against a trusted software composition analysis tool for integrity.

The central bank not only needs to strengthen its security measures around the infrastructure supporting the CBDC system but also take the lead to ensure the security posture for all the intermediary organizations, making sure they meet security requirements for performing the relevant CBDC functions. The Polaris framework for secure and resilience CBDC systems provides recommendations for central banks to achieve these objectives (BIS Innovation Hub, 2023d), including steps such as third-party attestation of their security posture. All organizations must establish sufficient security measures to prevent, detect, respond to, and recover from cybersecurity threats as listed earlier, as well as emerging threats and zero-day attacks.

Given that a CBDC needs to be designed to last for many years, the PKI infrastructure adopted by the central bank should be quantum ready (NIST, 2023). As part of a long-term plan, central banks may need to start by working with a PKI based on classic cryptography, but design and test a way for the CBDC system to work in a mixed-mode cryptography, in which some cryptography remains class, others migrated to post-quantum cryptography, and in some cases post-quantum cryptography need to apply on top of the classic ones, to ensure that baseline security is maintained even if the newer post-quantum cryptography has been broken by novel attacks (KTH, 2023) (BIS Innovation Hub, 2023c).

5.2.2 Security in Core CBDC Design

There are typically two ways to represent digital central bank money: token or account balance. When it comes to preventing double spending, counterfeiting, and replay attacks, the token-based approach can learn from the Bitcoin design in which the Unspent Transaction Output (UTXO) data structure recorded on the ledger provides an elegant design (Nakamoto, 2008), while the account balance approach can take hints from Ethereum, in which a state transition machine coupled with smart contracts are used to ensure the integrity of asset ownership (Antonopoulos & Wood, 2018).

In a token-based design, a CBDC token would have the following key components:

1. Face value—The token would have a designated monetary value, such as $1, $5, and $10, that is recognized as legal tender. This face value determines the worth of the token as a medium of exchange.
2. Unique identifier—Each token would have a unique ID number, serial code, or other identifier that distinguishes it from every other token generated. This allows for tokens to be tracked and accounted for on an individual basis.
3. Spent/unspent indication—The token would have a flag, switch, or other technical indicators showing whether this specific token has already been used in a transaction (spent) or not yet used (unspent). This prevents double spending of the same token.
4. Cryptographic proof of issuance—The token would contain a digital signature, certificate, or other cryptographic proof verifying that it has been legitimately generated by the central bank or an authorized intermediary. This validates the authenticity of the token.
5. Ownership indicator—Technically enforced rules embedded in the token would show what entity (person, business, etc.) has the right to use this token in a transaction. This could be in the form of digital keys, access controls, or terms of ownership. Transferring the token transfers ownership rights.

In Bitcoin, the above is implemented in the UTXO mechanism, which provides information needed for all required elements listed above, even though item 4 has been replaced somewhat by the consensus mechanism. A Bitcoin UTXO can only be claimed or spent by the party who can use their private key to generate the proof as required in item #5. Each Bitcoin transaction can have one or more UTXOs as input and one or more UTXOs as output.

In a design based on account balance, each account and its generated transactions have the following components:

1. Credentials for claiming ownership of the account (something you know, something you have, something you are, or a combination of the above).
2. Balance in the account.
3. The digital signature associated with a transaction that is signed by a private key that only the owner of the account can activate.
4. A unique identifier associated with the transaction.

Many central banks desire to support offline payments with their CBDCs, for reasons such as resilience and financial inclusion (BIS Innovation Hub, 2023b). Offline CBDCs have the challenge of not being able to rely on a central ledger in real time to detect double spending, replay, or other malicious attacks. Typically offline CBDC wallets are expected to connect online periodically, or it will no longer be usable after certain limits have been reached (e.g., number of payments and total amount that has been paid) in order to contain such risk.

Offline CBDCs have been designed to rely on the following components to minimize the possibility of such attacks being successful, and enable central banks to detect them at the earliest possible time:

1. During the enrollment process, a certificate is provisioned to the offline wallet. This certificate is based on a central bank-managed PKI to support the CBDC.
2. The offline wallet stores its cryptographic keys in a tamper-resistant hardware module.
3. The offline wallet software only runs in a secure environment such as a secure element or a trusted execution environment.
4. Each transaction generated by a wallet includes a unique nonce that is included as part of the transaction's digital signature generated by the wallet.
5. The wallet keeps track of all transactions it has generated or accepted and uploads these transactions to the online system when the wallet connects online.

Overall, to prevent and detect double spending, the security and integrity of the following components are key:

1. Private key of a CBDC wallet—The private key is the mathematically related cryptographic key that can authorize transactions from a specific wallet. Maintaining control and security of private keys is critical to prevent unauthorized transactions or double spending from a compromised wallet. Features like hardware security modules, multi-party security protocols, and key recovery options help secure private keys.
2. Wallet software and runtime environment—The wallet applications and computing systems supporting CBDC must utilize secure software and environments that resist hacking attempts to steal funds or spend them twice. This includes using encrypted data storage, code obfuscation, regular security audits, isolated runtimes, version tracking, and other protections against exploits or alterations of wallet programs.
3. CBDC ledger—An immutable, append-only ledger using distributed ledger technology underpins the recording of all CBDC transactions and token allocation. Its tamper-proof design preserves an accurate single record of token ownership and spending, while duplicate conflicting entries are blocked. Consensus mechanisms validate transactions. Off-chain transaction caching and settlement require coordination with the ledger to prevent double spending.
4. Timely detection of illicit transactions—Advanced analytics including machine learning algorithms applied to the CBDC ledger and transaction streams can identify statistical anomalies, suspicious activity patterns, and potential instances of fraud or abuse for further investigation. Near real-time detection of double spending attempts or other illegal activity enables rapid alerts and intervention to mitigate risks.

A compromise of one or more of the above provides a window of opportunity for an advisory to launch an attack for financial gains.

In summary, here are the most important security measures to implement for each scenario (Table 5.1).

Table 5.1 Summary of security measures in core CBDC design

	Online	Offline
Token-based design (UTXO)	• The recording of whether a token is spent or not must be trustworthy. • Each token must be uniquely identified. • The private key or program for claiming ownership of the token must be well protected.	• Each token must be uniquely identified and have a digital signature, verifiable by the CBDC PKI, indicating that it has been issued by the central bank or an authorized party with a trusted certificate. • Offline CBDC wallet must run in a trusted execution environment or secure element with assurance that the program logic has not been tampered with.
Account balance-based design	• The recording of whether a transaction is executed or not must be trustworthy. • Each transaction must be uniquely identified and digitally signed using the private key of the account owner or a program authorized by the account owner. • Each transaction must be validated for integrity on the server side (e.g., account has enough balance and transaction is not replayed) before initiation and execution. • Transaction execution (i.e., moving balance from one account to another) must be atomic so that there is no chance for a re-entrancy attack.	• Account balance must be stored in a trusted execution environment or secure element that is tamper resistant. • Offline wallet software must run in a trusted execution environment or secure element with assurance that the program logic has not been tampered with, and would assign a different transaction ID for each unique transaction. • Data for each transaction, including its unique identifier, must be signed using the wallet's private key. • The CBDC wallet's private key must be stored in tamper-resistant hardware, unless the wallet is only shortlived (e.g., no more than a few hours).
Both designs	• CBDC wallet software, as well as software that supports the CBDC lifecycle such as user enrollment, must be certified. • Ideally, a certified CBDC wallet software should store a digital certificate generated using the CBDC PKI. • The recipient's wallet could choose to only accept CBDC transactions from wallets that can present a valid certificate. • CBDC activities must be monitored to detect any potential anomalies, with established response and recovery measures.	• Each offline wallet must be registered, with its public key certified using the CBDC PKI, and with the certificate stored in the wallet. • Offline wallet must use its digital certificate to establish trust with the recipient's device. • For most solutions, limits must be established and enforced by the offline wallets so that they would not stay offline for too long. • Offline wallets must be able to enforce a block list so that they would not transact with wallets that are deemed compromised. • An anomaly detection and response mechanism must be in place to identify compromised wallets, add them to the block list, and populate the updated list to wallets when they connect online.

5.3   Central Bank's Role in Securing the CBDC

Monetary value carried by CBDCs are liability to the issuing central bank. In addition, any incidents to the CBDC system could cause damage to the reputation of the central bank. The stake is high for the central bank to make sure the CBDC system is reliable, secure, and resilient. As a system is as secure as its weakest link, the central bank needs to take a holistic approach to securing the CBDC ecosystem.

Within the central bank, the following areas need special attention in securing the CBDC:

- Security of Core Technology and Mining Process: The foundation of a secure CBDC lies in the integrity and robustness of its core technology. This includes the process of mining or generating CBDCs, which must be impervious to attacks or fraud. Ensuring cryptographic security, implementing rigorous testing protocols, and continuously updating the system to address emerging threats are vital steps in safeguarding this core process.
- Public Key Infrastructure (PKI): PKI plays a pivotal role in maintaining trust within the network of entities authorized to operate the CBDC system. It involves the use of digital certificates and cryptographic keys to authenticate the identity of these entities and secure communications. A robust PKI infrastructure is essential to prevent unauthorized access, ensuring that only verified entities can participate in the CBDC ecosystem.
- Distributed Ledger Technology (DLT) Management: If the CBDC employs DLT, the management of its permissioned network is crucial. This includes overseeing the membership of the network, defining and managing the roles of participants, and ensuring the life cycle of each participant is properly managed. Effective governance policies and procedures must be in place to ensure that the network operates seamlessly, securely, and in accordance with regulatory requirements.
- Distribution Technology and Processes: The method and technology used for distributing CBDCs to intermediaries are paramount. This involves not only the technical aspects of transferring digital currency but also ensuring that these processes are in compliance with regulatory standards, are transparent, and are secure from tampering or interception.
- Risk Management for CBDC Circulation: Monitoring the circulation of CBDCs is essential to identify any unusual activities or anomalies that could indicate potential threats or fraud. This requires a sophisticated risk management system capable of detecting, analyzing, and responding to these anomalies in real time. Such systems should incorporate advanced analytics and machine learning techniques to adapt to new patterns of misuse.
- Incident Response Framework: Finally, the central bank must establish a comprehensive framework for incident response. This encompasses the people, processes, and technology needed to quickly and effectively respond to security breaches or other incidents. A well-defined incident response plan ensures that

any disruptions to the CBDC system are managed and resolved swiftly, minimizing impact, and maintaining public trust in the digital currency.

Central banks must also play a role in fostering cyber resilience in the overall CBDC ecosystem, both within the jurisdiction and across borders. Such efforts may need to be carried out with other authorities, and could include:

- Cyber Resilience Readiness Assessments: Central banks should lead in assessing the preparedness of the CBDC ecosystem to handle cyber threats. This involves conducting regular evaluations, supervising activities, and sharing knowledge and best practices. Through these assessments, potential vulnerabilities can be identified and addressed, ensuring that the CBDC infrastructure is robust against cyberattacks.
- Security Standards for Ecosystem Participants: Setting clear and stringent security standards is crucial for all participants in the CBDC ecosystem, including intermediaries like wallet providers and technology infrastructure firms. These standards should encompass aspects like data protection, transaction security, and network integrity. Compliance with these standards ensures a uniform level of security across all entities involved in the CBDC framework.
- Security Certification Program: Central banks could establish a security certification program for all organizations participating in the CBDC ecosystem. The certification process would vary based on the risk profile associated with each organization's role in the CBDC operations. This certification ensures that all participants meet the required security benchmarks, thereby enhancing the overall trustworthiness of the CBDC system.
- Cybersecurity Preparedness Exercises: Organizing preparedness exercises such as table-top simulations and cyber range exercises with participating organizations is essential. These exercises provide a practical platform for testing the resilience of the CBDC system against potential cyber threats and also help in identifying areas for improvement. Regular drills ensure that participants are well-prepared to respond effectively in the event of a real cyber incident.
- Threat Intelligence Sharing: Establishing a mechanism for sharing threat intelligence related to the CBDC is vital for a proactive defense strategy. This involves disseminating information about potential cyber threats and vulnerabilities both within the jurisdiction and internationally. By sharing intelligence, central banks and other participants can stay ahead of emerging threats and coordinate their defensive measures more effectively.
- Coordinated Incident Response: Central banks must spearhead the development of a coordinated incident response strategy for the CBDC ecosystem. This strategy should outline clear protocols for responding to cyber incidents, including the roles and responsibilities of different stakeholders. A coordinated response ensures swift and effective action in the event of a cyberattack, minimizing the impact on the CBDC network and maintaining public confidence.

5.4 CBDC and Blockchain Security

Among the central banks that have launched CBDCs, some have adopted distributed ledger technologies (DLT), including the Central Bank of Nigeria and the Eastern Caribbean Central Bank (The Atlantic Council, 2021). It is conceivable that more central banks will choose DLT/blockchain as the core technology for their CBDC solutions.

In addition, DLT has shown some promise in wholesale use cases and has been adopted by some large financial institutions for interbank or intrabank settlements (Huan, 2023). In cases where DLT would be used, such platforms may require customized security architecture and additional hardening as compared to traditional systems (Chain Security: Nodes, Algorithm, and Network, 2023). This would also require actors in a CBDC ecosystem to develop, attract, and retain new talent required to design, implement, and operate secure and resilient DLT systems. The large-value attacks on DLT protocols and smart contracts in the DeFi space underscore the potential operational and reputational risks. A study of six DeFi hacks by the BIS (BIS Innovation Hub, 2023e) showed that DLT-based solutions may be vulnerable to some new attack techniques, including:

- Exploiting Authorization Weakness with Brute Force: Attackers might target smart contracts by brute-forcing a 32-bit hash value of the smart contract ID. Since a 32-bit hash offers limited combinations, it is vulnerable to brute-force attacks. Upon successfully guessing the hash, attackers could insert their public key into the list of authorized entities. This breach would allow them to illegitimately transfer or withdraw large amounts of assets from the network.
- Inserting Attacker's Public Key: The strategy of inserting an attacker's public key into a list of authorized entities is a direct consequence of the successful brute-force attack on the smart contract's ID hash. This action effectively grants the attacker the same privileges as legitimate entities, enabling them to transfer assets out of the network without alerting other participants until after the theft has occurred.
- Compromising Nodes for Consensus Control: By compromising a majority of the nodes in a blockchain network, attackers can potentially take control of the consensus mechanism. This allows them to approve malicious transactions or even propose and enact changes to the network that facilitate asset theft. Such a situation is particularly dangerous in networks where consensus is achieved democratically, as it undermines the very foundation of trust and security in blockchain technology.
- Manipulation via Flash Loans: Utilizing flash loans to momentarily gain substantial voting rights in a proof-of-stake system is another sophisticated attack. The attacker borrows a large amount of cryptocurrency through a flash loan, which is a loan taken and repaid within a single transaction block. This temporarily inflated stake can be used to influence network decisions or changes, potentially to the detriment of the network's security and integrity.
- Developer Credential Compromise: If an attacker gains access to a developer's credentials, they can inject malicious code into a smart contract. This could result

in various malicious activities, such as creating backdoors, redirecting funds, or altering the contract's logic to facilitate fraud.

- Exploiting Unlimited Approval Mechanisms: Some blockchain systems allow users to give a smart contract permission to transfer an unlimited amount of a certain asset on their behalf. Attackers might exploit this by tricking users into approving malicious contracts, which can then drain the user's assets without requiring further permissions.
- Altering Transaction Signature Verification: Replacing the default program responsible for verifying transaction signatures with one controlled by the attacker could allow unauthorized transactions. This breach would enable the attacker to forge transactions or bypass security checks, undermining the integrity of the entire blockchain system.
- Re-entrancy Attacks on Smart Contracts: This attack involves exploiting a re-entrancy vulnerability in a smart contract. An attacker can repeatedly withdraw funds in a single transaction before the contract's balance is updated. This type of vulnerability was famously exploited in the DAO attack on the Ethereum network, leading to significant financial losses.

These exploits are often unfamiliar to information security teams and need to be understood, with countermeasures developed and embedded into the security processes and tools used by organizations in the CBDC ecosystem.

In addition, a Feds Notes article on CBDC security, with the assumption of a DLT-based architecture, highlighted areas that are unique for DLTs (Hansen & Delak, 2022), including:

- Risk of Adopting Novel Technology: The adoption of DLT, being a relatively new technological paradigm, inherently carries certain risks. These include uncertainties in scalability, performance under different stress scenarios, and the technology's long-term viability. As DLT continues to evolve, there is also the challenge of adapting to emerging standards and practices, which may not be fully established or widely accepted yet.
- Weakness in DLT Governance Models: The governance model of a DLT system is crucial for its successful operation. Weaknesses in governance can lead to inefficiencies, conflicts, or vulnerabilities in decision-making processes. This includes how changes to the protocol are proposed, reviewed, and implemented, and how disputes are resolved. Ensuring a transparent, equitable, and effective governance model is critical for the stability and security of the DLT-based CBDC system.
- Risks with Cryptographic Key Custodianship: In a DLT-based CBDC system, the security of cryptographic keys is paramount, as these keys control access to funds. Risks arise from how these keys are stored, managed, and protected by custodians. The potential for loss, theft, or compromise of these keys poses a significant risk to the security of the CBDC. Implementing robust key management practices and secure storage solutions is essential to mitigate these risks.
- Integrity of External Oracles: External oracles are third-party services that feed external data into the blockchain. The integrity of these oracles is crucial, as they can be a source of vulnerability if the data provided is incorrect or manipulated.

Ensuring the reliability and security of these oracles is essential to maintain the accuracy and trustworthiness of the data used in smart contracts and other DLT operations.

- Threats from Quantum Computing: The advent of quantum computing presents a potential threat to the cryptographic algorithms currently used in DLT systems. Quantum computers, with their advanced computational capabilities, could theoretically break the cryptographic codes that secure DLT networks. Proactively researching and developing quantum-resistant cryptographic algorithms is vital to safeguard the future security of the CBDC.
- Vulnerabilities in Software Code: The software code underlying the DLT and its applications, such as smart contracts, can contain vulnerabilities or bugs that could be exploited by attackers. Continuous code reviews, rigorous testing, and audits are necessary to identify and rectify any such vulnerabilities. It is also crucial to establish a protocol for swiftly addressing any discovered issues to maintain the integrity and security of the system.

Countermeasures for such risks are enumerated in the chapter on security for permissioned blockchain, as included in Volume 1 of this book (Zhang, 2023).

5.5   CBDC and Financial Crime

Like other payment methods, CBDC is likely to be leveraged by financial criminals to conduct illegal activities. Aside from double spending and counterfeiting of CBDCs that have been discussed before, criminals could leverage CBDCs to launder money or commit financial fraud, as evidenced by some recent cases during the eCNY pilot in China (Shen et al., 2022). Likewise, cryptocurrencies such as Bitcoin, with its pseudonymous nature and privacy services such as mixers, have been leveraged by cybercriminals for laundering stolen cryptocurrencies, which could have been purchased with illegal fiat money (CertiK, 2023).

For crime prevention, financial institutions need to perform due diligence checks including know-your-customer (KYC), anti-money laundering (AML), and counter finance of terrorism (CFT) before onboarding a user to the CBDC platform, unless the central bank allows certain types of CBDC wallets to bypass such check under the condition of limitation in services (Ledger Insights, 2021). With its adoption of digital technologies, CBDC systems can leverage more advanced techniques including machine learning to achieve better effectiveness and efficiency. Project Aurora from the BIS demonstrated how machine learning and privacy enhancing technologies, especially graph neural networks and federated learning, can be used together to effectively fight financial crimes while at the same time preserving user privacy (BIS Innovation Hub, 2023a).

As of now, detection of financial crime activities in the payment systems is the responsibility of commercial banks, who typically focus on meeting the compliance requirements, which may not lead to the best results in capturing illicit activities.

Central banks could use the launch of its CBDC as an opportunity to bring improvements in this area. For example, the CBDC solution could be accompanied by a financial crime detection service that is powered by advanced analytics and machine learning model, which can differentiate between normal transactions vs potential criminal activities based on existing machine learning models that are fine tuned by recent transaction data. In cross-border scenarios, having a joint facility that takes data from the different parties to generate alerts and share intelligence with all participants could be an approach to be explored (Marr, 2023).

There is a trade-off between user privacy and financial crime prevention and detection. Having commercial banks conduct these checks avoids the concentration of sensitive user data at the central bank. In addition, different privacy-enhancing technologies can be applied in this process, as well as in situations when data from multiple commercial banks need to be aggregated to detect illicit activities across financial institutions, and even across jurisdictions. The applicability of such technologies in the context of CBDC will be discussed in the next section.

5.6 CBDC and Privacy

Privacy protection is a major requirement for many CBDC projects (Ahnert et al., 2022). Given that CBDC is a form of digital money that will be used for digital payments, it is technically very hard for CBDCs to provide the same level of privacy as cash. Even though a central bank may allow CBDC wallets that are not linked to any user identity, with limits on the amount of CBDCs that can be held and used, a user's digital footprint associated with using the CBDC wallet or the device and communications network that the CBDC wallet runs on may give away data that can be used to trace back to the identity of the CBDC user (Mitnick & Vamosi, 2019).

Ways to enable anonymity when using CBDCs include:

- Prepaid CBDC cards that can be given to and used by anyone with basic authentication such as a PIN code. The card is not associated with any personal identity and thus maintains the anonymity of the payer.
- eCash2.0, as demonstrated in BIS project Tourbillon (BIS Innovation Hub, 2023f), uses the blind signature technique to prove the validity of a CBDC token without revealing information of its owner (Chaum & Moser, 2022).

Privacy concerns with digital payments are not unique to CBDCs. However, given that a CBDC potentially could give the central bank access to detailed financial data of individual citizens, and the perception that central banks are connected to other parts of the government that have more power over citizens than any commercial entities who may also have a large amount of data about citizens, people tend to be more worried about the privacy issue related to CBDCs than with services offered by private entities. These concerns could be addressed by institutional design of the overall CBDC system (Uhlig et al., 2023). For example, in the two-tier

CBDC design, users would interact with financial intermediaries that have provided them the CBDC wallets. The central bank does not have access to the identity information of users, which is kept by the financial intermediaries. Access to such data by the central bank or other government agencies would be subject to the same legal frameworks as other digital payments. Central banks could use aggregated data, with differential privacy safeguards (Dwork & Roth, 2014), in order to receive timely data to help it perform its duties in a digital economy.

In addition, a CBDC system should consider adopting privacy-enhancing technologies (PETs) where appropriate (Darbha & Arora, 2020; OECD, 2023). For example,

- Data obfuscation tools such as zero-knowledge proofs (ZKPs) could potentially allow transaction verification without revealing the identity of the parties involved or the amount being paid. Anonymization and pseudonymization techniques would focus on hiding the identity of the payer and payee in the transaction record.
- Data encryption is a tool that should be applied where possible to protect private data. This includes data encryption in transit using transport layer security, data encryption at rest using symmetric encryption algorithms such as 256-bit advanced encryption system (AES) technology, and data encryption using homomorphic encryption (HE), multi-party computation (MPC), etc. Given that HE and MPC are still emerging and require a dramatic increase in the amount of computation, and may not be feasible for many use cases, the trusted execution environment is often used to achieve data protection in use.
- Federated and distributed analytics, such as federated learning, enable analytics, and discovery to be carried out without providing personal data to another party. These arrangements need to be validated to make sure there is no data leakage in the process (Boernert et al., 2023).
- Data accountability tools focus on organization-level controls on who can have access to privacy-sensitive data and potentially give controls to data owners to grant access after validation of purpose and qualification.

Besides technical measures such as the device level privacy options and the application of PETs as mentioned above, consumer awareness and education is another key component for privacy protection. For example, the effort by the Bank of England to have an extensive consultation with its citizens about the CBDC sets a good example (Bank of England & HM Treasury, 2023). Users of CBDCs should be made aware of the different options and their privacy implications. As demonstrated in a recent study of London's Oyster cards for its public transportation system, commuters often go for convenience over privacy protection, knowingly or otherwise (Walker, 2024). Of course, data privacy protection must reconcile with compliance requirements associated with payments. Full disclosure of such requirements as well as the different CBDC choices and their privacy implications should be part of each central bank's CBDC rollout plan.

5.7   Conclusion

Retail CBDCs bring a number of new challenges to central banks. From the security perspective, besides the general cybersecurity practices that apply to all use cases, the following areas have emerged as requiring special attention in designing a CBDC solution:

- Novel technologies bring novel attacks. As observed in recent hacks in the DeFi space, attackers have been able to devise new attack techniques before they were recognized by the design team or the cybersecurity function. In other cases, known vulnerabilities such as the re-entrancy vulnerability in smart contracts have become well known to the blockchain community but have not been universally integrated into the security process or toolbox for development teams, which translates into repeated occurrences that have caused significant financial losses.
- New usage scenarios introduce new challenges that require careful design to address. One example is offline CBDC payment as desired by many central banks. Given the amount of time the attacker may have trying to compromise the cryptographic key or the program execution environment on an offline payment device, the fact that malicious transactions can only be detected at a later time when the involved devices connect online, central banks need to make their design decisions carefully in order to strike a balance between usability (as related to their policy objectives for offline payments) and the prevention and detection of illicit transactions.
- Potential issues associated with a complex ecosystem. As central banks likely need to work with the private sector for distributing CBDCs, including onboarding users, a security weakness in these institutions could pose a significant risk to the integrity of the overall CBDC system. Initial and ongoing assessments of their security readiness and the ongoing monitoring of the overall CBDC system are crucial for maintaining its integrity.
- Supply chain compromises pose a significant threat to CBDC systems. From infrastructure dependencies such as a cloud service provider, a software update process, to software dependencies such as a code library that the development team has adopted, a security breach of such components could have a significant impact on the CBDC system. Central banks must establish a robust dependency management program to mitigate such risks.
- The durability requirement means the need to prepare for known and unknown attacks. A CBDC needs to be designed to last for many years. The threat from quantum computing on today's cryptographic algorithms is a known risk that is already on the horizon. How to design agility into the CBDC solution so that it can replace a vulnerability component (such as a critical cryptographic algorithm) with a new one without requiring a total overhaul of the solution is an important task. This is an area that central banks could learn from private sector companies that have a track record of operating a durable system that has gone through many changes over the years.

Given all these considerations, central banks need to have cybersecurity as a top priority in their CBDC projects, with clear requirements that must be met for each step of their CBDC journey, especially for going into pilot and production. They must also establish strong monitoring, detection, and incident response capabilities, regularly re-assess the threat and risk landscape, and identify the areas that require the CBDC to adapt and evolve.

References

Ahnert, T., Hoffmann, P., & Monnet, C. (2022, December). The digital economy, privacy, and CBDC. European Central Bank.

Antonopoulos, A. M., & Wood, G. (2018). Mastering Ethereum: Building Smart Contracts and DApps. O'Reilly.

Atlantic Council. (2021, July 21). Central Bank Digital Currency Tracker. Atlantic Council. Retrieved January 20, 2024, from https://www.atlanticcouncil.org/cbdctracker/

Atlantic Council. (2022, June). Missing Key: The Challenge of Cybersecurity and Central Bank Digital Currency.

Auer, R., & Böhme, R. (2020, March). The technology of retail central bank digital currency. BIS Quarterly Review.

Bank of England & HM Treasury. (2023, February 7). The digital pound: A new form of money for households and businesses? Bank of England. Retrieved January 20, 2024, from https://www.bankofengland.co.uk/paper/2023/the-digital-pound-consultation-paper

BIS Innovation Hub. (2023a, May). Project Aurora - The power of data, technology and collaboration to combat money laundering across institutions and borders.

BIS Innovation Hub. (2023b, May). Project Polaris - Part 1: A handbook for offline payments with CBDCv.

BIS Innovation Hub. (2023c, June). Project Leap - Quantum-proofing the financial system.

BIS Innovation Hub. (2023d, July). Project Polaris Part 2: A security and resilience framework for CBDC systems.

BIS Innovation Hub. (2023e, July). Project Polaris: Part 3: Closing the CBDC cyber threat modelling gaps.

BIS Innovation Hub. (2023f, November). Project Tourbillon: Exploring privacy, security and scalability for CBDCs.

BIS Representative Office for the Americas. (2023, November). CBDC information security and operational risks to central banks.

Boernert, E., Chmiel, J., & Antczak, L. (2023, September 28). Preventing Health Data Leaks with Federated Learning Using NVIDIA FLARE | NVIDIA Technical Blog. NVIDIA Developer. Retrieved January 23, 2024, from https://developer.nvidia.com/blog/preventing-health-data-leaks-with-federated-learning-using-nvidia-flare/

CertiK. (2023, December 17). Dirty Laundry: The Bitcoin Network's Growing Role in the Laundering of Stolen Crypto - Blog - Web3 Security Leaderboard. CertiK. Retrieved January 20, 2024, from https://www.certik.com/resources/blog/28KkDOQrEF1D0ef6lP2vLX-dirty-laundry-the-bitcoin-networks-growing-role-in-the-laundering-of-stolen

CertiK. (2024, January 3). Hack3d: The Web3 Security Report 2023 - Blog - Web3 Security Leaderboard. CertiK. Retrieved January 20, 2024, from https://www.certik.com/resources/blog/7BokMhPUgffqEvyvXgHNaq-hack3d-the-web3-security-report-2023

Chain Security: Nodes, Algorithm, and Network. (2023). In K. Huang, D. Budorin, L. J. Tan, W. Ma, & Z. W. Zhang (Eds.), A Comprehensive Guide for Web3 Security: From Technology, Economic and Legal Aspects (pp. 31-60). Springer Nature Switzerland. https://doi.org/10.1007/978-3-031-39288-7

Chaum, D., & Moser, T. (2022). eCash 2.0 - Inalienably private and quantum-resistant to counterfeiting.

Cockroach Labs. (2020). Transport Layer Security (TLS) and Public Key Infrastructure (PKI). Cockroach Labs. Retrieved January 20, 2024, from https://www.cockroachlabs.com/docs/stable/security-reference/transport-layer-security

Darbha, S., & Arora, R. (2020, June). Privacy in CBDC technology. Bank of Canada. Retrieved January 23, 2024, from https://www.bankofcanada.ca/2020/06/staff-analytical-note-2020-9/

The Digital Dollar Project. (2023, July). Secure Adoption of a Digital Dollar - Operational and Compliance Risks for the U.S. Banking Sector.

Dwork, C., & Roth, A. (2014). The Algorithmic Foundations of Differential Privacy. Now Publishers.

Hansen, T., & Delak, K. (2022, February 3). Security Considerations for a Central Bank Digital Currency. Board of Governors of the Federal Reserve System. Retrieved January 16, 2024, from https://www.federalreserve.gov/econres/notes/feds-notes/security-considerations-for-a-central-bank-digital-currency-20220203.html

Huang, C. (2023, November 15). MAS-led Project Guardian adds five more pilots in asset tokenisation. The Straits Times. Retrieved January 20, 2024, from https://www.straitstimes.com/business/mas-led-project-guardian-adds-five-more-pilots-in-asset-tokenisation

KTH. (2023, April 18). Researchers found leak in cryptographic algorithm. KTH. Retrieved January 20, 2024, from https://www.kth.se/en/eecs/nyheter/kth-forskare-lyckades-hitta-lacka-i-krypto-grafisk-algoritm-1.1248786

Ledger Insights. (2021, June 11). Details about the digital yuan wallet officially disclosed. Ledger Insights. Retrieved January 20, 2024, from https://www.ledgerinsights.com/details-about-the-digital-yuan-wallet-officially-disclosed/

Marr, B. (2023). The Amazing Ways How Mastercard Uses Artificial Intelligence To Stop Fraud And Reduce False Declines. Bernard Marr. Retrieved January 20, 2024, from https://bernard-marr.com/the-amazing-ways-how-mastercard-uses-artificial-intelligence-to-stop-fraud-and-reduce-false-declines/

Mitnick, K., & Vamosi, R. (2019). The Art of Invisibility: The World's Most Famous Hacker Teaches You How to Be Safe in the Age of Big Brother and Big Data. Little, Brown.

Nakamoto, S. (2008, October). Bitcoin: A Peer-to-Peer Electronic Cash System.

NIST. (2023, December). Migration to Post-Quantum Cryptography Quantum Readiness: Cryptographic Discovery.

OECD. (2023, March). Emerging Privacy Enhancing Technologies: Current Regulatory and Policy Approaches.

Security in Permissioned Blockchain. (2023). In K. Huang, D. Budorin, L. J. Tan, W. Ma, & Z. W. Zhang (Eds.), A Comprehensive Guide for Web3 Security: From Technology, Economic and Legal Aspects. Springer Nature Switzerland.

Shen, T., McMahon, K., & Shah, M. (2022, September 27). China busts US$28 mln digital yuan money laundering case. Forkast News. Retrieved January 20, 2024, from https://forkast.news/china-busts-digital-yuan-money-laundering-case/

Syed, M., & Lee, A. (2022, April 22). Designing the IT Solution Architecture for a Central Bank Digital Currency (CBDC). LinkedIn. Retrieved January 20, 2024, from https://www.linkedin.com/pulse/designing-solution-architecture-central-bank-digital-currency-syed/

Uhlig, H., Alonso, M., & Frost, J. (2023). Privacy in Digital Payments—Escaping the Panopticon. Georgetown Journal of International Affairs, 24(2), 174-180. https://doi.org/10.1353/gia.2023.a913643

US Department of Defense. (2021, March). DoD Enterprise DevSecOps Reference Design.

Walker, M. C. (2024, January 2). What London's Oyster cards reveal about central bank digital currencies. LSE Blogs. Retrieved January 15, 2024, from https://blogs.lse.ac.uk/businessreview/2024/01/02/what-londons-oyster-cards-reveal-about-central-bank-digital-currencies/

Zhijun William Zhang   is the Technology and Innovation Adviser at the Bank for International Settlements (BIS) Innovation Hub—Nordic Centre, where he focuses on cybersecurity and resilience for future financial market infrastructure. Before joining the BIS, he was the lead security architect at the World Bank Group, where his team was responsible for security architecture design and assessment of all technology platforms and business solutions. He also led the security and risk work for WBG's innovation lab. Before joining the WBG, William worked at The Vanguard Group in various capacities, including user experience design, emerging technology research, system architecture, and information security. William received his BS degree from Peking University, and his PhD from the University of Maryland, both in computer science.

Part II
Frontiers of Web3 Security

While Part 1 established the security foundations underpinning much of Web3 applications today, the chapters in this next part examine futuristic domains pushing the boundaries of possibility at the intersection of blockchain and security.

Chapter 6 explores the promise and current limitations of utilizing Web3 to tackle the global ransomware epidemic. Ransomware attacks have paralyzed corporations, hospitals, cities, and critical infrastructure. Blockchains' tamper-proof records and decentralized infrastructure offer enhanced resilience. Smart contracts allow automatic responses and payments upon attack detection. However, adoption barriers around integration costs, performance limits, and regulatory unclearness persist. A nuanced roadmap balancing pragmatic solutions with innovative potential is needed.

Securing increasingly complex technology supply chains is an escalating challenge. Chapter 7 analyzes risks unique to Web3 based supply chains like dependence on decentralized service providers and risks from interconnected code dependencies. Auditing procedures, SBOMs, and blockchain-enabled provenance tracking can enhance integrity. Regulatory initiatives around critical software transparency and disclosure will likely expand to cover Web3 ecosystems.

While AI solutions, especially the much hyped Generative AI solutions, can profoundly amplify the effectiveness of Web3 security controls, the integration risks from improperly secured AI must also be addressed. Chapter 8 examines governance frameworks around rigorous testing, validation, and monitoring to ensure reliable and safe AI utilization. Perspectives on utilizing Web3 to also secure AI systems against data poisoning, model theft, and manipulation attacks are discussed. The relationships between AI progress and Web3 will continue to co-evolve.

Chapter 9 is dedicated to assessing and defending against potentially devastating impacts of quantum computing on Web3 in the next decade. Current cryptographic schemes face risk from rapid quantum algorithm advances. Migration to quantum-resistant cryptography, quantum key distribution, and hybrid multi-layered solutions with classical encryption, blockchain consensus protocols, and hardware security modules is imperative.

Lastly, privacy preservation in Web3 systems involves complex technical trade-offs between utility and confidentiality across layers—applications, smart contracts, consensus protocols, and network messaging. Chapter 10 offers a comprehensive guide to techniques that mask sensitive data including zero knowledge proofs, homomorphic and multiparty encryption. It provides frameworks on architecture decisions for decentralized data sharing while respecting privacy.

The five chapters in Part 2 analyze bleeding edge topics that while seemingly futuristic today, will likely become mainstream security considerations for Web3 in the near future. Mastering these new frontiers requires interdisciplinary expertise across blockchain, security, cryptography, AI safety, hardware systems and regulation. Part 2 equips readers with perspectives spanning both opportunities and risks of integrating innovations across these domains with Web3.

We invite you to explore with us on the frontiers of Web3 security in the chapters that follow. Mastering these domains demands intellectual rigor and an openness to imagine possibilities balanced with pragmatism. May the perspectives presented spur discussion, debate, and further collaborative progress at the intersection of blockchain and security.

Chapter 6
Web3 and Ransomware Attacks

Jerry Huang ⓘ and Ken Huang ⓘ

Abstract This chapter explores the escalating threat of ransomware attacks and the potential of blockchain technology as a defense mechanism. Ransomware, known for encrypting victims' files and demanding ransoms, presents a significant threat across various sectors. This chapter examines blockchain's strengths and weaknesses in combating ransomware, focusing on its immutability, data protection, and decentralized storage capabilities. It discusses the technical architecture of blockchain in resisting ransomware, including data entry, chain linkage, distributed ledgers, and consensus mechanisms. The chapter also addresses implementation considerations like smart contract programming, data access control, and integration with external threat intelligence for automated incident response.

Ransomware attacks have surged in recent times, with prominent targets spanning government agencies, sports teams, IT companies, and publishers. These malicious acts involve encrypting a victim's files and demanding ransoms for their release, typically executed via malware. This chapter delves into a potential defense mechanism against such threats: blockchain technology. While blockchain holds promise in mitigating ransomware attacks, it is not without its shortcomings. We aim to provide a balanced perspective on the strengths and weaknesses of blockchain in the fight against ransomware.

J. Huang
The University of Chicago, Chicago, IL, USA
e-mail: jerryh@uchicago.edu

K. Huang (✉)
DistributedApps LLC, Fairfax, Virginia, USA
e-mail: ken@distributedapps.ai

6.1 Ransomware Landscape and Prominent Attacks

In today's rapidly evolving digital landscape, being proactive in seeking solutions is vital. Yet, to truly address the challenges posed by cyber threats, especially ransomware, we must first comprehensively understand the nature of the beast we are up against. Ransomware, a particularly malicious form of cyberattack, has evolved over the years to become a significant menace to both individuals and organizations. Its primary modus operandi involves infiltrating systems, encrypting valuable data, and then demanding a ransom from the victims to restore access to their own information.

TechTarget, a respected name in the realm of IT information, has extensively documented some of the most devastating ransomware attacks in recent history. Their reports shed light not just on the sophistication of these attacks, but also on their far-reaching consequences (Waldman 2023). From crippling essential infrastructure, causing financial losses running into millions, to bringing operations of multinational corporations to a standstill, the repercussions of these attacks are both diverse and profound.

One might wonder, why is understanding these specific incidents crucial? The answer lies in the patterns and tactics employed by the attackers. By dissecting these high-profile incidents, we can glean insights into the vulnerabilities that were exploited, the sectors that were targeted, and the evolution of ransomware strains. These insights are invaluable as they guide us in identifying potential weak points in our own systems and fortifying them accordingly.

Furthermore, an appreciation of the scale of these attacks underscores the urgent need for robust countermeasures. This is not merely about preventing financial losses or operational disruptions. At stake is the very trust that users place in digital systems. Each successful ransomware attack erodes this trust, making it imperative for stakeholders across the board—from tech developers to policymakers—to collaborate and devise effective strategies to combat this menace.

6.1.1 Anatomy of a Ransomware Attack

Ransomware, over the years, has solidified its position as a formidable cyber threat. Distinct from other forms of cyber aggression, ransomware brazenly holds a user's data hostage. It leverages encryption as its primary weapon, locking out users and demanding a ransom for the data's release. To devise effective countermeasures against ransomware, one must first delve deep into its operational intricacies, tracing its path from initial infiltration to the final ransom demand.

As shown in Fig. 6.1, the genesis of a ransomware attack lies in its initial infiltration. Cybercriminals employ a gamut of tactics to gain unauthorized entry into target systems. Among the most prevalent is the use of phishing emails. Cleverly masquerading as legitimate communications, these emails ensnare unsuspecting users into downloading malevolent attachments or guiding them to malicious websites. Parallelly, exploit kits play a significant role. These sophisticated tools

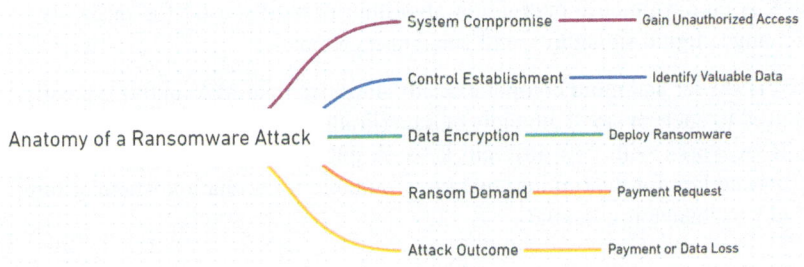

System Compromise ——— Gain Unauthorized Access

Control Establishment ——— Identify Valuable Data

Anatomy of a Ransomware Attack ——— Data Encryption ——— Deploy Ransomware

Ransom Demand ——— Payment Request

Attack Outcome ——— Payment or Data Loss

Fig. 6.1 Anatomy of a ransomware attack

continuously scan systems, hunting for vulnerabilities. Upon identifying a weak spot, these kits spring into action, exploiting the vulnerability to plant the ransomware. Additionally, the threat of drive-by downloads looms large. In such scenarios, individuals, by merely visiting a compromised website, inadvertently initiate the automatic download of ransomware.

Once the malicious software has nestled itself within a system, it does not always manifest its intent immediately. In many instances, the ransomware bides its time, surreptitiously moving laterally across networks. During this phase, it actively seeks out valuable data troves or aims to compromise an extensive array of systems. Such a strategy is tactical, as it amplifies the ensuing encryption's impact, thereby augmenting the potential ransom sum.

As the ransomware concludes its reconnaissance, it transitions to its hallmark stage: encryption. Here, the malware encrypts the victim's files, effectively locking them out. Some advanced variants elevate their malevolence by tampering with boot records, rendering entire systems defunct. The encryption algorithms employed, such as RSA or AES, are formidable. When executed proficiently, the encrypted data becomes virtually irretrievable without the unique decryption key.

AES (Advanced Encryption Standard) and RSA (Rivest, Shamir, Adleman) are both encryption algorithms used to secure data, but they serve different purposes and have different characteristics.

AES is a symmetric encryption algorithm designed for rapid data encryption and decryption.

– It is commonly used for encrypting large amounts of data, such as in file storage and streaming services, due to its fast encryption and decryption speeds (Poggi 2021).
– AES uses the same secret key for both encryption and decryption, requiring secure key distribution to all parties involved in the communication.

RSA is an asymmetric encryption algorithm primarily used for secure key exchange, digital signatures, and email encryption.

- It is slower and more computationally intensive than AES, and is typically used to encrypt small amounts of data (Franklin and Tozzi 2022).
- RSA works with two different keys: A public key for encryption and a private key for decryption, making it suitable for scenarios where secure key exchange is essential.

Following the encryption onslaught, victims are confronted with the ransomware's ultimatum, typically in the form of a ransom note. This communication elucidates the encryption specifics and delineates the ransom payment modalities. Cryptocurrencies, particularly Bitcoin, are the preferred payment mode, given their potential to mask the attacker's identity. These notes often inject a sense of urgency by imposing a deadline, post which the ransom might escalate or, in more dire scenarios, the decryption key might be permanently obliterated.

In the event the victim acquiesces to the ransom demand, a treacherous path lies ahead. While some cybercriminals might uphold their end of the bargain, releasing the decryption key upon payment, others might renege. Entrusting one's fate to the whims of cybercriminals is a gamble, with no guarantees of data retrieval. Moreover, capitulating to their demands could embolden them, marking the victim for future exploitations.

Concluding a ransomware attack heralds a phase of recovery and reflection. Victims grapple with system rehabilitation, which encompasses purging the system of malware remnants, restoring data from backups, and enhancing security protocols. This post-attack phase underscores the indelible impact of ransomware, necessitating a renewed emphasis on proactive security measures and user awareness.

6.1.2 High-Profile Ransomware Incidents

The ubiquity of ransomware attacks over the past few years has been startling. Not confined to any particular sector or region, these attacks have transcended boundaries, targeting entities ranging from municipal bodies to global corporations. By examining some of the most impactful incidents, as documented by TechTarget, we can gain a deeper understanding of the ransomware threat landscape.

One of the most significant disruptions occurred when government systems were targeted. Municipal services were halted, and critical data was held hostage. An illustrative case is the attack on the city of Atlanta in 2018. The SamSam ransomware infiltrated the city's systems, crippling numerous services, from bill payments to court proceedings (Perlroth 2018). The attackers demanded a ransom in Bitcoin, while the city grappled with restoring its systems. The incident not only resulted in financial losses but also highlighted the vulnerabilities in governmental digital infrastructures.

The world of sports was not immune either. In a surprising turn of events, several sports teams found their data encrypted and operations hindered. A notable incident involved a major European football club (Palmer 2020). The turnstiles of an English football club were locked due to a ransomware attack, nearly leading to the cancellation of a league fixture. The attack crippled corporate and security systems, preventing fans from entering or leaving the stadium, and could have cost the club hundreds of thousands of pounds in lost income. This incident is part of a broader trend of cybercriminals targeting sports organizations, with the UK's National Cyber Security Centre reporting that over 70% of sports institutions have been victims of cyberattacks, including phishing, ransomware, and other forms of cybercrime. This incident was a stark reminder that even sectors seemingly unrelated to the digital realm could be vulnerable to such attacks.

IT companies, which one might assume would be fortified against such threats, have also fallen victim. In a paradoxical twist, those who provide digital solutions were themselves ensnared by ransomware. A Russian-speaking hacker group, Clop, has targeted several IT industry companies, including IBM, Cognizant, and Deloitte, with data extortion attempts. The attacks exploited a vulnerability in Progress' MOVEit file transfer software, leading to a spate of data breaches that have affected the IT channel. Additionally, the hacker group behind the MOVEit attacks has been observed exploiting a now-patched vulnerability in the SysAid IT service management platform, a competitor to ServiceNow and Jira (Alspach 2023). These attacks serve as a reminder that even businesses dedicated to cybersecurity are not immune to the threat posed by cybercriminals.

Lastly, the publishing sector became an inadvertent target. News agencies and media houses found their operations halted as ransomware encrypted their files. For example, the Guardian suffered a ransomware attack in December 2022 that impacted parts of its technology infrastructure. The "highly sophisticated" attack likely started with a phishing attempt that allowed unauthorized third-party access. Personal data of UK staff was compromised, including sensitive information like identification documents and salaries, but there was no evidence that reader or subscriber data was impacted. The attack disrupted some backend services and forced remote work. The company worked with law enforcement and outside experts to recover affected systems (Waterson 2022).

These incidents, while just a snapshot of the myriad of attacks, underscore the pervasive and indiscriminate nature of ransomware. From sectors steeped in technology to those on its periphery, no entity seems impervious. The audacity and scale of these attacks underscore the urgent need for robust cybersecurity measures and heightened awareness across all sectors.

The ransomware landscape saw continued growth and disruption in both 2022 and 2023. Victims spanned multiple industries, with a particular impact on technology, healthcare, education, and manufacturing sectors.

Table 6.1 gives the summary of the major ransomware attacks in 2022 and 2023 based on the information provided in the reports Waldman (2023) and Gihon (2023):

Table 6.1 Major ransomware attacks in 2022 and 2023

Date	Ransomware group	Victim	Sector	Impact
Feb 2022	BlackByte	San Francisco 49ers	Sports	Attack limited to corporate network
May 2022	Quantum	Glenn County Office of Education	Education	Paid $400 K ransom
May 2022	Unknown	Opus Interactive	Technology	Customer workloads restored
May 2022	Yanluowang	Cisco	Technology	No ransomware deployed, data stolen
June 2022	LockBit	Entrust Corporation	Technology	Data exfiltrated
June 2022	Unknown	Macmillan Publishers	Publishing	Operations disrupted
Sept 2022	Vice Society	Los Angeles Unified School District	Education	Data leaked online
Oct 2022	Unknown	CommonSpirit Health	Healthcare	Patient data breached
Nov 2022	Unknown	Apprentice Information Systems	Technology	Services for 31 counties disrupted
Dec 2022	Play	Rackspace	Technology	Hosted Exchange service disrupted
Jan 2023	LockBit3.0	Royal Mail	Logistics	£65 million ransom demanded
Jan 2023	Nevada	ESXi servers	Technology	Mass campaign, thousands infected
Feb 2023	Hive	Multiple victims	Multiple	Operation dismantled by FBI
Q1 2023	Clop	Multiple victims	Multiple	Large victim count increase
Q2 2023	MalasLocker	Multiple victims	Multiple	171 victims in debut quarter
Q2 2023	8Base	Multiple victims	Multiple	107 victims in debut quarter

6.1.3 The Cost of Ransomware

When a ransomware attack strikes, the ramifications extend well beyond the initial payment demanded by the attackers. The aftermath of such an intrusion is multifaceted, with costs spiraling in directions that are not immediately apparent. These costs, whether monetary, operational, or reputational, underscore the profound impact ransomware can have on its targets.

At a fundamental level, the direct financial implications are evident. Victims are often faced with a stark choice: pay the demanded ransom in the hope of regaining access to their data or bear the costs of restoring systems and recovering data from backups, if available. While the ransom amount can vary, ranging from a few hundred dollars for individual targets to millions for large corporations, it is just the tip of the iceberg. For businesses, the downtime resulting from a ransomware attack can lead to substantial revenue losses, especially for those whose operations rely heavily on real-time data access.

Moreover, the process of recovery is neither straightforward nor cheap. Organizations often need to engage cybersecurity experts to cleanse their systems, restore data, and bolster their defenses. These remediation efforts, combined with potential hardware replacements and software upgrades, can escalate costs exponentially. For those without recent backups, the price of data reconstruction, if at all feasible, can be exorbitant.

However, not all costs are quantifiable. The intangible repercussions of a ransomware attack can be even more debilitating. An organization's reputation, painstakingly built over years, can be tarnished overnight. Clients, stakeholders, and partners might question the organization's competence and its commitment to data security. Such doubts can lead to loss of business, contract terminations, or legal actions. For public companies, stock prices might plummet, reflecting the market's shaken confidence.

Furthermore, there is a human cost to consider. Employees might face increased workloads as they scramble to address the fallout, leading to stress and burnout. In some instances, businesses, unable to bear the financial strain, might resort to layoffs or even face bankruptcy.

On a broader scale, significant ransomware attacks, especially those targeting critical infrastructure, can have ripple effects on economies. For instance, an attack on a major utility provider can disrupt services for millions, while an intrusion into a financial institution can shake the very foundations of economic trust.

To put it into concrete perspective with numbers. The average cost to remediate a ransomware attack is around $1.85 million, and the ransom payment is just a small part of the total cost, estimated to be around 15% (Blosil 2022). The true cost of a ransomware attack extends far beyond the ransom payment and can add up to be seven times the ransom demand. This includes lost revenue, costs of bringing in contractors, legal fees, and the impact on the organization's reputation (SpyCloud 2023). Rebuilding a damaged reputation is a challenging and time-consuming process, requiring transparent communication, proactive security improvements, and a commitment to regaining trust. The reputational damage can lead to loss of customers and clients, affecting the organization's long-term viability and competitiveness in the market (Ton 2022). According to Cybersecurity Ventures, global ransomware damage costs are predicted to exceed $265 billion by 2031, and the financial damage of ransomware is only part of the picture; it can also cause reputational and operational damage (Osborne 2023). Therefore, it is essential for organizations to consider the full impact of ransomware attacks, including the costs associated with reputation damage, when evaluating their cybersecurity strategies and risk management approaches.

6.2   Blockchain as a Defense Mechanism

Blockchain technology, underpinning the Web3 revolution, offers mechanisms that could potentially mitigate ransomware threats. This section delves into the features of blockchain that make it a potential defensive tool against ransomware and how organizations can leverage it.

6.2.1 Immutable Records and Data Protection

Blockchain's architecture is designed fundamentally around the principle of immutability. Once data is entered into a blockchain, it becomes part of a "block," which is subsequently added to a chain of existing blocks. Crucially, each block contains a cryptographic hash of the preceding block, creating a tightly interlinked chain. This linkage means that any attempt to alter data in a block would not only require changes to the hash of that block but also cascade changes throughout the subsequent blocks in the chain. Given that blockchains are distributed ledgers, with multiple copies existing across various nodes, achieving such a widespread, consistent alteration is computationally prohibitive, if not impossible.

In the context of ransomware, this immutability becomes a significant deterrent. Ransomware thrives on its ability to encrypt data, rendering it inaccessible to the user. However, data stored on a blockchain is resistant to such encryption due to its immutable nature. Even if a cybercriminal were to compromise a node or a user's access point, the underlying data on the blockchain remains unaltered. Moreover, since the data is replicated across multiple nodes in the network, recovery becomes a more straightforward process. Rather than paying a ransom to decrypt data, users can rely on the integrity of the blockchain to retrieve unaltered data from another node.

Furthermore, blockchain's transparency, another consequence of its distributed nature, ensures that any unauthorized access or suspicious activity is quickly detected. Each transaction on the blockchain is visible to all participants, and any attempt to alter or encrypt data would be immediately evident, allowing for swift countermeasures.

Nevertheless, as discussed in Sect. 6.3.1, it is not advisable to backup all data to blockchain due to scalability issues, using a decentralized data storage approach is better (Sect. 6.2.2). So, we advise only storing necessary and critical data on the chain with strong encryption and also storing the hash of a collection of data on the chain for integrity tracking.

Another interesting approach proposed by some researchers is to utilize blockchain technology for recording the pre- and post-encryption behavior of ransomware attacks, with a specific focus on crypto ransomware (Singh and Ali 2022). The challenge highlighted by the researcher is the current lack of precise data regarding the behavior of ransomware attacks both before and after encryption. This deficiency impedes the effective development of detection models, as accurately selecting features and identifying similarities in attacks are crucial for designing effective defense mechanisms. The researcher's approach aims to bridge this gap by creating a blockchain-based repository of attack data, which could offer a more comprehensive and secure record. Conventionally, strategies like regular file backups are recommended as countermeasures against ransomware. However, the researcher rightly notes that such measures are not without significant drawbacks. Regular backups, particularly key backup schemes, can entail substantial computational resources, thereby incurring high costs. Moreover, these strategies are more reactive, offering solutions after an attack has occurred, rather than proactively preventing or mitigating it. The researcher's proposal to use blockchain technology aims not just to create a detailed and secure log of ransomware behaviors but also to potentially uncover patterns and similarities that could aid in early detection and prevention of such attacks. This proactive stance could considerably enhance the effectiveness of cybersecurity measures against ransomware, especially sophisticated variants like crypto ransomware.

The architecture and implementation of using immutable records to defend against ransomware is discussed below.

6.2.1.1   Technical Architecture

In the context of defending against ransomware, the technical architecture of blockchain technology offers several key features that make it inherently resistant to such attacks (Fig. 6.2):

1. Data Entry and Block Creation:
   - When data is recorded in a blockchain, it is encapsulated within a block. This block contains a cryptographic hash, which is a unique identifier for the block's contents. In the event of a ransomware attack, the attackers often seek to encrypt data for ransom. However, in a blockchain, because each block's contents are hashed, any unauthorized alteration (like encryption by ransomware) would change the hash, instantly signaling a breach.

2. Chain Linkage and Immutability:
   - The blockchain's structure, where each block contains the hash of the previous block, creates a chain of dependencies. This interlinked chain means that altering a single block (as ransomware might attempt) would require recalculating the hashes of all subsequent blocks. The computational effort to achieve this, especially on a blockchain of significant size, is virtually impossible, thus providing a strong defense against ransomware that seeks to alter or encrypt data.

3. Distributed Ledger and Network Nodes:
   - Blockchain's distributed ledger architecture means that the data is not stored in a single location but is replicated across multiple nodes. In the case of a ransomware attack, even if one node is compromised, the other nodes in the network maintain intact copies of the blockchain. This redundancy ensures that the data remains accessible and unaltered, thwarting the typical ransomware strategy of locking users out of their data.

4. Consensus Mechanisms:
   - Consensus algorithms are employed to maintain a consistent state across the blockchain. These mechanisms ensure that any new entry (such as a block) is validated by multiple nodes. In the context of ransomware defense, this means any unauthorized changes (like those attempted by ransomware) would fail to achieve network-wide consensus and therefore would not be recorded on the blockchain.

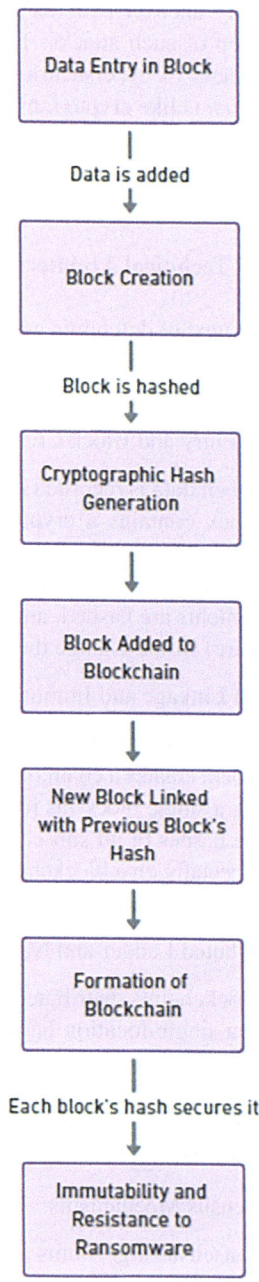

Fig. 6.2 Blockchain immutable records as ransomware defense

5. Transaction Visibility and Network Transparency:

  – Blockchain's transparent nature ensures that all transactions are visible to
    network participants. This visibility is crucial in detecting and responding
    to ransomware attacks. If ransomware were to attempt to alter data, such
    changes would be visible to all participants, enabling rapid detection and
    response.

This technical architecture underpins the robustness of blockchain technology, making it an ideal platform for applications requiring high security, transparency, and data integrity. The combined use of cryptographic hashing, chain
linkage, distributed ledger technology, consensus mechanisms, and network
transparency creates an environment where data is both immutable and readily
verifiable, establishing blockchain as a formidable tool against data tampering
and cyber threats.

6.2.1.2 Implementation Considerations

The following are considerations when implementing the solution:

Data Recovery
In the event of a ransomware attack, blockchain technology enables effective data
recovery. Since the blockchain is a distributed ledger, with each node in the network holding a copy of the entire blockchain, the data is replicated across multiple points. This means that if one node is compromised, the data can still be
retrieved unaltered from other nodes. This capability significantly reduces the
efficacy of ransomware attacks, which typically rely on denying access to data.
Users can bypass the need to pay a ransom by accessing their data from other
nodes in the blockchain network.

Selective Data Storage
While blockchain offers significant advantages in data protection, it is crucial to
consider its scalability limitations. Storing large volumes of data directly on the
blockchain can lead to inefficiencies. To address this, it is advisable to store only
critical data on the blockchain. For larger datasets, storing hashes of data collections
on the blockchain can be a practical approach. These hashes can be used for integrity tracking and verification, ensuring data has not been tampered with, while the
bulk of the data is stored off-chain.

6.2.1.3 Code Example

The following is a snippet of a smart contract code to handle the storage of important encrypted data on-chain, stores hashes of data sets, and provide mechanisms for
data recovery and validation in the event of ransomware.

```solidity
1.  // SPDX-License-Identifier: MIT
2.  pragma solidity ^0.8.23;
3.
4.  contract DataProtection {
5.      // Mapping to store important data on-chain
6.      mapping(address => string) private importantData;
7.
8.      // Mapping to store the hash of off-chain data
9.      mapping(bytes32 => bool) private offChainDataHashes;
10.
11.     // Event to indicate data storage
12.     event ImportantDataStored(address indexed user, string
    data);
13.     event DataHashSet(bytes32 dataHash);
14.
15.     // Function to store important data on-chain, the data must
    be encrypted prior to storage
16.     function storeImportantData(string memory data) public {
17.         importantData[msg.sender] = data;
18.         emit ImportantDataStored(msg.sender, data);
19.     }
20.
21.     // Function to store the hash of off-chain data
22.     function storeDataHash(bytes32 dataHash) public {
23.         require(!offChainDataHashes[dataHash], "Data hash
    already exists.");
24.         offChainDataHashes[dataHash] = true;
25.         emit DataHashSet(dataHash);
26.     }
27.
28.     // Function to retrieve important data in case of an event
    like ransomware
29.     function recoverImportantData(address user) public view
    returns (string memory) {
30.         require(bytes(importantData[user]).length > 0, "No data
    stored for this user.");
31.         return importantData[user];
32.     }
33.
34.     // Function to validate the hash of off-chain data
35.     function validateDataHash(bytes32 dataHash) public view
    returns (bool) {
36.         return offChainDataHashes[dataHash];
37.     }
38. }
```

Let us see how this code works:

1. Storing Important Data On-Chain:

   – "storeImportantData": This function allows users to store critical data directly on the blockchain. The data must be encrypted before being stored. The encryption should be done off-chain and is not shown in the code example. This data is linked to the user's address and can be anything deemed important enough to warrant on-chain storage.

2. Storing Hashes of Off-Chain Data:

   – "storeDataHash": Since storing large amounts of data on-chain is not scalable, this function enables users to store the hash of a data set that resides off-chain. This hash acts as proof of the data's integrity.

3. Recovering Important Data:

   – "recoverImportantData": In the event of a ransomware attack, users can call this function to retrieve their important data stored on-chain.

4. Validating Off-Chain Data:

   – "validateDataHash": This function allows users to verify the integrity of their off-chain data. By providing the hash of the data set, users can check if it matches the hash stored on-chain, ensuring the data has not been tampered with.

Integration with Off-Chain Storage
– The actual off-chain data should be stored in a secure, decentralized storage system like IPFS (InterPlanetary File System) or a similar distributed database (see Sect. 6.2.2).
– When data is added to this off-chain storage, its cryptographic hash is calculated and then stored on the blockchain using the "storeDataHash" function.
– To validate the integrity of this data, retrieve it from the off-chain storage, compute its hash, and use the "validateDataHash" function to ensure it matches the hash stored on the blockchain.

This example illustrates a basic implementation. Depending on the specific requirements and security considerations, additional features such as access controls, encryption, and more sophisticated data handling mechanisms might be necessary.

6.2.2 Decentralized Data Storage

Decentralization stands at the core of the blockchain ethos. It signifies a move away from centralized systems, where data is stored at a singular or limited number of points, to a model where data is distributed across a multitude of nodes. This unique structure has profound implications for data security, especially when considering the threat posed by ransomware.

In a conventional centralized system, data storage often resembles a fortified castle—a single, concentrated repository of information protected by layers of defenses. While this might seem secure, it also represents a singular target, a single point of failure. If ransomware attackers breach this fortress, the entirety of the data is at their mercy. This central vulnerability is what many ransomware attacks exploit, encrypting vast swaths of data in one fell swoop.

Enter blockchain's decentralized storage mechanism. Rather than having data housed in one location, it is dispersed across a network of nodes, each holding a copy of the entire dataset. This dispersion fundamentally alters the attack dynamics for ransomware perpetrators. Instead of targeting a central repository, they would need to compromise a majority of nodes in the network to exert any meaningful control over the data. Given the global and distributed nature of many blockchain networks, this is a task of Herculean proportions.

Additionally, even if a node or a subset of nodes were compromised, the decentralized nature of the blockchain ensures data integrity is maintained. Since each node possesses a full copy of the blockchain, data can easily be retrieved and restored from uncompromised parts of the network. This redundancy acts as a bulwark against data loss and provides a level of resilience that centralized systems often cannot match.

Beyond mere data retrieval, decentralized storage also offers proactive protection against ransomware. Given that ransomware operates by encrypting data and demanding a ransom for its decryption, the distributed nature of data on a blockchain makes it considerably more challenging for ransomware to execute a successful encryption across the network. Even if a few nodes get affected, the broader network remains untouched, ensuring continued access to unencrypted data.

Implementing a decentralized storage solution as a defense against ransomware involves a few key components: the technical architecture, implementation details, and coding examples. Let us break down each of these components (also shown in Fig. 6.3):

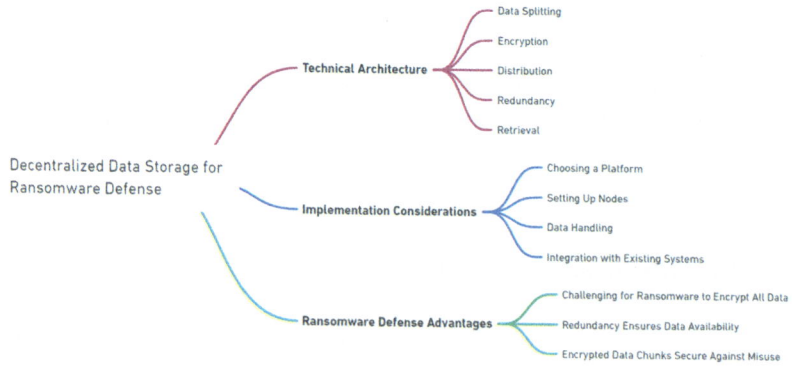

Fig. 6.3 Decentralized data storage for ransomware defense

6.2.2.1   Technical Architecture

Decentralized storage systems, such as IPFS (InterPlanetary File System, see Box below) and blockchain-based storage solutions like Storj (Jacobi 2023) and Filecoin (Hussey 2020), represent a paradigm shift in how data is stored and managed.

IPFS (InterPlanetary File System) is a protocol and peer-to-peer network for storing and sharing data in a distributed file system (McKay 2022). Some key points about IPFS:

- It is decentralized—Data is stored across a distributed network of nodes rather than in centralized servers. This makes it more resilient and harder to censor.
- It uses content addressing—Files are given a unique fingerprint called a cryptographic hash instead of location-based addressing. This identifies content by what it is rather than where it is stored.
- It is fast and efficient—By allowing nodes to transfer data directly without central servers, it can achieve high transfer speeds and efficiency.
- Data is resilient—Multiple nodes store copies of files, so if one node goes offline the data is still available. This leads to high resilience against failure or attack.
- It has versioning built-in—Files are immutable, so updating a file creates a new version while keeping the old version still accessible. This preserves history.

In summary, IPFS provides a decentralized storage and sharing network for resilient, efficient, and verifiable distribution of content. It complements or could potentially replace existing Internet infrastructure for a more open and resilient web.

Unlike traditional centralized storage systems where data is stored in a single location or server, decentralized storage spreads data across a vast network of nodes. This architectural choice significantly bolsters data resilience and provides a robust defense against ransomware attacks.

1. Data splitting: The first step in a decentralized storage process is data splitting. In this phase, large files are broken down into smaller, manageable pieces. This division is not just for ease of storage but also for enhancing security and efficiency in data handling. When data is split into smaller chunks, it reduces the risk of complete data compromise, as ransomware would need to corrupt many pieces across different locations to affect the entire file.
2. Encryption: Once the data is divided, each piece is encrypted. This encryption is a critical security measure, adding an additional layer of protection. By encrypt-

ing each chunk of data separately, decentralized storage systems ensure that even if some parts of the data are intercepted or accessed without authorization, they remain indecipherable and useless to the attacker. This layer of encryption is particularly vital in safeguarding sensitive information against ransomware, which often seeks to exploit unencrypted or weakly encrypted data.

3. Distribution: After encryption, these data chunks are distributed across a network of nodes. In decentralized systems like IPFS or blockchain-based platforms, these nodes can be spread globally. This wide distribution means that no single point of failure exists. If one node is compromised or goes offline, the data remains accessible through other nodes. Such distribution not only enhances data availability but also significantly mitigates the risk of total data loss due to localized ransomware attacks.

4. Redundancy: A key feature of decentralized storage is redundancy. Data is not just stored in multiple places; it is replicated across various nodes. This redundancy ensures that even if some nodes are affected by a cyberattack, other nodes can compensate, ensuring data availability and integrity. In the context of defending against ransomware, redundancy means that even if some parts of the network are held hostage by ransomware, the overall integrity and accessibility of the data remain intact.

5. Retrieval: Retrieval in decentralized storage involves accessing these distributed and replicated nodes to reassemble the data chunks. When a user requests data, the system locates the various pieces across its network, decrypts them, and reassembles them into the original file. This process is typically transparent to the user, offering a seamless experience similar to traditional storage systems.

The decentralized approach offers several advantages in the context of ransomware defense. Firstly, the distributed nature of data storage makes it exceedingly challenging for ransomware to locate and encrypt all copies of the data. Secondly, the inherent redundancy in the system means that even if some data is compromised, unaffected copies remain available for restoration. Lastly, the encryption of each data chunk ensures that even if ransomware accesses some data, it cannot decrypt or misuse it without the requisite encryption keys.

6.2.2.2 Implementation Considerations

Implementing a decentralized storage solution involves several crucial steps, each contributing to the overall effectiveness and security of the system. The implementation process can be outlined in the following stages:

1. Choosing a Platform: The first step in implementing a decentralized storage solution is selecting an appropriate platform. Popular options include IPFS, Storj, and Filecoin. Each of these platforms has unique features and operational mechanisms. For instance, IPFS offers a peer-to-peer hypermedia protocol designed to make the web faster, safer, and more open. Storj, on the other hand, emphasizes privacy and security, providing decentralized cloud storage solutions. Filecoin operates as a decentralized storage network based on blockchain technology. The choice of plat-

form should be influenced by specific needs such as data volume, security requirements, cost considerations, and the desired level of decentralization.

2. Setting Up Nodes: Once the platform is chosen, the next step is setting up nodes that will participate in the storage network. Nodes are individual storage points within the network, and they can be either on-premises servers or cloud instances. The decision between on-premises and cloud-based nodes depends on factors like budget, available infrastructure, and data governance policies. On-premises nodes offer more control and potentially better security, while cloud instances provide scalability and ease of management. When setting up these nodes, it is essential to ensure they meet the platform's technical requirements in terms of storage capacity, network connectivity, and computational power.

3. Data Handling: Data handling in decentralized storage systems involves more than just storing and retrieving data. Before uploading data to the network, it should be encrypted. This encryption serves as the first line of defense against unauthorized access and ransomware attacks. Managing encryption keys is also a critical aspect of data handling. Secure key management practices must be in place to prevent unauthorized access to the encryption keys. Key management involves securely storing, rotating, and disposing of keys according to best practices and compliance requirements.

4. Integration with Existing Systems: Integration of the decentralized storage solution with existing systems is crucial for its successful deployment. This integration should be seamless, allowing users to access and backup data as they would with traditional storage solutions. Integration involves configuring the decentralized storage to work with existing databases, applications, and backup systems. It may also require custom development work to ensure compatibility and optimal performance. For instance, APIs provided by the decentralized storage platform can be used to integrate it with existing applications, allowing for automated data backups and retrievals.

Each of these steps plays a vital role in the effective implementation of a decentralized storage solution. Choosing the right platform ensures that the system meets the specific needs of the organization. Setting up nodes correctly is crucial for the reliability and performance of the network. Proper data handling, especially regarding encryption and key management, is essential for security. Finally, seamless integration with existing systems ensures that the decentralized storage solution enhances, rather than disrupts, existing workflows. By carefully considering and executing each of these steps, organizations can leverage decentralized storage solutions to enhance their data storage and security strategies, particularly in the context of defending against ransomware attacks.

6.2.2.3 Code Example

Here is a simple example using IPFS with JavaScript. This example assumes you have an IPFS node running and "ipfs-http-client" installed. The code demonstrates how to use encryption to enhance the security of data stored on IPFS. It ensures that data is encrypted before leaving the client and remains encrypted while stored on IPFS nodes. Only someone with the correct decryption key (in this case, the

ENCRYPTION_KEY) can decrypt and access the original data. This approach adds
a layer of security, making the data resilient against unauthorized access and making the solution suitable for sensitive data storage.

```
1. const ipfsClient = require('ipfs-http-client');
2. const crypto = require('crypto');
3.
4. // Replace these with your actual encryption key and algorithm
5. const ENCRYPTION_KEY = 'your-encryption-key'; // Must be 256 bits (32
   characters)
6. const ALGORITHM = 'aes-256-cbc';
7.
8. // Connect to an IPFS node using HTTPS
9. const ipfs = ipfsClient.create({ host: 'ipfs.infura.io', port:
   '5001', protocol: 'https' });
10.
11.     // Function to encrypt data
12.     function encrypt(text) {
13.         let iv = crypto.randomBytes(16);
14.         let cipher = crypto.createCipheriv(ALGORITHM,
   Buffer.from(ENCRYPTION_KEY), iv);
15.         let encrypted = cipher.update(text);
16.         encrypted = Buffer.concat([encrypted, cipher.final()]);
17.         return iv.toString('hex') + ':' + encrypted.toString('hex');
18.     }
19.
20.     // Function to decrypt data
21.     function decrypt(text) {
22.         let textParts = text.split(':');
23.         let iv = Buffer.from(textParts.shift(), 'hex');
24.         let encryptedText = Buffer.from(textParts.join(':'), 'hex');
25.         let decipher = crypto.createDecipheriv(ALGORITHM,
   Buffer.from(ENCRYPTION_KEY), iv);
26.         let decrypted = decipher.update(encryptedText);
27.         decrypted = Buffer.concat([decrypted, decipher.final()]);
28.         return decrypted.toString();
29.     }
30.
31.     // Function to upload encrypted data to IPFS
32.     async function uploadToIPFS(data) {
33.         const encryptedData = encrypt(data);
34.         const { path } = await ipfs.add(encryptedData);
35.         return path;
36.     }
37.
38.     // Function to download and decrypt data from IPFS
39.     async function downloadFromIPFS(path) {
```

```
40.            const chunks = [];
41.            for await (const chunk of ipfs.cat(path)) {
42.                chunks.push(chunk);
43.            }
44.            const encryptedData = Buffer.concat(chunks).toString();
45.            return decrypt(encryptedData);
46.        }
47.
48.    // Example usage
49.    (async () => {
50.        const data = 'Hello, decentralized world!';
51.        const path = await uploadToIPFS(data);
52.        const downloadedData = await downloadFromIPFS(path);
53.        console.log(downloadedData); // Outputs: Hello,
    decentralized world!
54.    })();
```

Implementing decentralized storage requires careful planning and consideration of security, scalability, and compliance aspects. The code examples provided are basic and would need to be adapted and expanded for real-world applications.

6.2.3 Smart Contracts and Automated Responses

Smart contracts, sometimes referred to as the "business logic" of the blockchain, have been heralded for their potential to revolutionize various sectors, from finance to supply chains. Essentially self-executing contracts with the terms of the agreement directly written into lines of code, they can operate without intermediaries, ensuring that predefined criteria, once met, automatically trigger specific actions. When viewed through the lens of ransomware defense, smart contracts open up a new frontier of automated responses that could be pivotal in thwarting cyber threats.

First, smart contracts can be programmed to monitor data access patterns. Ransomware often exhibits telltale signs when it infiltrates a system—rapid file access, multiple encryption requests, or unusual data traffic. A smart contract, attuned to these irregularities, can detect these aberrant patterns in real time. Upon detection, it can trigger automated responses such as halting certain processes, isolating affected nodes, or even initiating backup protocols to preserve data integrity.

Furthermore, smart contracts can serve as guardians of critical data access permissions. By setting stringent criteria for data access and modification, smart contracts can act as gatekeepers, ensuring that only legitimate requests are processed. If a ransomware attempts to encrypt data, the smart contract could deny the operation unless it meets the predefined criteria, effectively stonewalling the attack.

Another potential application lies in integrating smart contracts with external threat intelligence platforms. These platforms continually update their databases with the latest ransomware signatures and attack vectors. A smart contract, in sync with such a platform, could automatically update its defense parameters, ensuring that the system remains fortified against the latest threats.

Lastly, in the unfortunate event of a successful ransomware attack, smart contracts can play a role in the aftermath. They can be designed to automatically notify stakeholders, initiate incident response procedures, or even liaise with cybersecurity firms to address the breach. This swift, automated response can significantly reduce the downtime and mitigate the overall impact of the attack.

Figure 6.4 shows a high-level diagram of how smart contracts work in ransomware defense.

Implementing a smart contract-based system to monitor and respond to ransomware threats involves several steps, from designing the architecture to writing and deploying the smart contracts. Let us break down the process:

6.2.3.1   Technical Architecture

1. Blockchain Platform: The smart contracts can be deployed on the Ethereum blockchain. Ethereum is chosen for its robust smart contract capabilities and widespread adoption.
2. Off-Chain Data Access Pattern Monitoring: Due to Ethereum's limitations in handling large volumes of data and real-time processing, an off-chain system will monitor data access patterns. This system will communicate with the smart contracts.
3. Integration with External Threat Intelligence Platforms: The system will integrate with external threat intelligence platforms to obtain the latest ransomware signatures and attack vectors.
4. Automated Response Mechanism: The smart contracts will be programmed to execute automated responses upon detecting ransomware activity.
5. User Interface (UI): A Web3 UI will be developed to interact with the smart contracts, providing a dashboard for monitoring and control.

6.2.3.2   Implementation Considerations

When implementing a smart contract-based solution to defend against ransomware, several key considerations must be taken.

Firstly, the smart contract, which forms the core of this system, needs to be adeptly programmed. Its primary function would be to identify patterns that resemble ransomware activity. This detection would rely on data access patterns that are reported by an off-chain monitoring system. This approach requires the smart contract to be intricately coded to discern normal access behaviors from those indicative of ransomware attacks.

Fig. 6.4 Smart contact for
ransomware defense

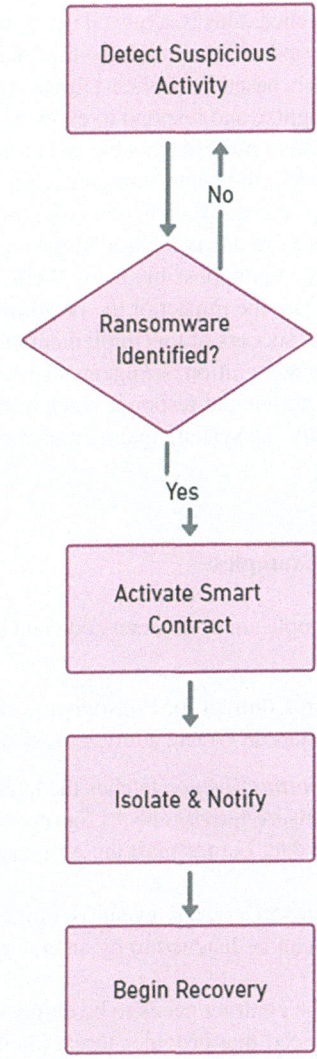

Additionally, the smart contract must incorporate robust data access controls. It should be designed to automatically deny any operations that appear suspicious or do not conform to the predefined criteria of legitimate data access. This control mechanism is vital in preventing unauthorized data modifications or access, which are common in ransomware attacks.

Another critical feature involves the integration of the smart contract with external threat intelligence. This is achieved using an Oracle mechanism, which serves as a bridge to relay updated threat data to the smart contract. By regularly updating its defense parameters based on the latest threat intelligence, the smart contract remains equipped to recognize and respond to evolving ransomware tactics.

Lastly, the system must be capable of automated incident response. In the event of a detected attack, the smart contract is programmed to take immediate actions such as notifying relevant stakeholders and triggering predefined incident response procedures. This response is facilitated through the emission of events that can be captured and acted upon, possibly using Websockets, to ensure a swift and effective response to mitigate the impact of the ransomware attack.

In essence, the success of this implementation hinges on a sophisticated blend of proactive pattern recognition, stringent access controls, dynamic threat intelligence integration, and automated response mechanisms. This comprehensive approach is designed to fortify the system against ransomware intrusions, ensuring data integrity and security.

6.2.3.3 Code Examples

We provide a sample smart contract code and also a snippet of UI code for the proposed solution:

1. Solidity Smart Contract for Ransomware Detection and Response. The contract constructor takes an Oracle address to set up the data feed.

   getLatestThreatIntelligence fetches the latest data from the oracle.

   detectRansomwareActivity checks for unauthorized access and a high threat level from the Oracle data before marking an attack and emitting the RansomwareDetected event.

   The *RansomwareDetected* event is emitted when a ransomware activity is detected, which can be listened to by an external WebSocket server or other listening services.

   Remember, the contract needs to be deployed with the address of an appropriate Chainlink Data Feed that provides threat intelligence data. The threshold for what constitutes a threat in detectRansomwareActivity (someThresholdValue) should be set based on the specifics of the threat intelligence data format.

```solidity
1.      pragma solidity ^0.8.23;
2.  // Import Chainlink interfaces
3.  import
    "@chainlink/contracts/src/v0.8/interfaces/AggregatorV3Interface.sol";
4.
5.  contract RansomwareDefense {
6.      // Define variables for storing data access patterns and threat
    intelligence
7.      mapping(address => bool) private legitimateAccess;
8.      address[] public affectedNodes;
9.      bool public isAttackDetected = false;
10.
11.         // Oracle data feeder interface
12.         AggregatorV3Interface internal threatIntelligenceFeed;
13.
14.         // Event definitions
15.         event RansomwareDetected(address indexed node);
16.
17.         // Constructor to set the oracle data feeder address
18.         constructor(address _oracleAddress) {
19.             threatIntelligenceFeed =
    AggregatorV3Interface(_oracleAddress);
20.         }
21.
22.         // Function to update data access patterns (called by off-
    chain monitoring system)
23.         function updateDataAccessPattern(address node, bool
    isLegitimate) external {
24.             legitimateAccess[node] = isLegitimate;
25.         }
26.
27.         // Function to get the latest threat intelligence data
28.         function getLatestThreatIntelligence() public view returns
    (int256) {
29.             (,int256 answer,,,) =
    threatIntelligenceFeed.latestRoundData();
30.             return answer;
31.         }
32.
33.         // Function to detect ransomware activity
34.         function detectRansomwareActivity() external {
35.             int256 threatLevel = getLatestThreatIntelligence();
36.             if (!legitimateAccess[msg.sender] || threatLevel >
    someThresholdValue) {
37.                 isAttackDetected = true;
38.                 affectedNodes.push(msg.sender);
```

```
39.              emit RansomwareDetected(msg.sender); // Emitting an
    event
40.                  // Trigger automated response
41.              handleAttack();
42.          }
43.      }
44.
45.      // Function to handle detected ransomware attack
46.      function handleAttack() internal {
47.          // Logic to dispatch events to isolate affected nodes,
    notify stakeholders, etc.
48.      }
49.  }
```

2. Web3 UI and Wallet Connect Code

For the Web3 UI, you would use React Native to create a user interface that interacts with the Ethereum blockchain.

```
1. import Web3 from 'web3';
2. import { abi } from './RansomwareDefense.json'; // Import ABI of the
   updated smart contract
3.
4. const web3 = new Web3(Web3.givenProvider || "ws://localhost:8545");
5. const contractAddress = 'YOUR_CONTRACT_ADDRESS';
6. const ransomwareDefenseContract = new web3.eth.Contract(abi,
   contractAddress);
7.
8. // Function to get the latest threat intelligence data
9. const getLatestThreatIntelligence = async () => {
10.       const threatLevel = await
   ransomwareDefenseContract.methods.getLatestThreatIntelligence().call(
   );
11.       console.log("Latest Threat Intelligence: ", threatLevel);
12.       return threatLevel;
13.   };
14.
15.   // Function to call smart contract's detectRansomwareActivity
16.   const detectRansomwareActivity = async () => {
17.       const accounts = await web3.eth.getAccounts();
18.
   ransomwareDefenseContract.methods.detectRansomwareActivity().send({
   from: accounts[0] })
19.           .on('receipt', receipt => {
20.               console.log("Ransomware detection transaction
   receipt: ", receipt);
21.           });
22.   };
23.
24.   // Listening for the RansomwareDetected event
25.   ransomwareDefenseContract.events.RansomwareDetected({
26.       fromBlock: 'latest'
27.   }, function(error, event) {
28.       if (error) console.error(error);
29.       else console.log("Ransomware Detected Event: ", event);
30.   });
31.
32.   // UI Component to trigger detection and display threat
   intelligence
33.       // ... React Native component code
```

In this code:

getLatestThreatIntelligence Function: This function communicates with the smart contract to fetch the latest threat intelligence data.

detectRansomwareActivity Function: This function calls the detectRansomware-Activity method of the smart contract. It handles the transaction receipt for any state changes in the blockchain.

Event Listener: The code listens for the *RansomwareDetected* event using ransomwareDefenseContract.events.RansomwareDetected. When the event is detected, it logs the event details.

Web3 Provider: The *Web3.givenProvider* is used if available (e.g., MetaMask in a browser environment); otherwise, it falls back to a local provider.

React Native UI Components: The actual UI components in React Native would include elements to display the threat intelligence data and provide a user interface to trigger the ransomware detection function. The specifics of these components depend on your application's design and user experience requirements.

Remember to replace *"YOUR_CONTRACT_ADDRESS"* with the actual deployed address of your *RansomwareDefense* smart contract. Additionally, ensure that the ABI (RansomwareDefense.json) is updated to match the latest version of your smart contract.

6.2.4 Efforts in Combating Ransomware by US Government Agencies

The research conducted by American University outlines various strategies and actions taken by the US government to combat ransomware. A key approach is the use of blockchain technology and cryptocurrency regulation as both defensive and offensive tools (American University 2023).

1. Defensive Use of Blockchain and Cryptocurrency Regulation: The USA has adopted an "all tools" approach to improving critical infrastructure cybersecurity. This includes new regulations for natural gas and oil pipelines and the Cybersecurity and Infrastructure Security Agency's (CISA) "Shields Up" campaign. The Department of Justice has also engaged in various initiatives such as StopRansomware.gov (Department of Justice 2021), Civil Cyber Fraud Initiative, and National Cryptocurrency Enforcement Team to dismantle the cyber infrastructure used by criminals.
2. Improving Visibility into Cryptocurrency Transactions: The Office of Foreign Assets Control (OFAC) and the Financial Crimes Enforcement Network (FINCEN) are working to enhance transparency into cryptocurrency offramps, where criminals convert illicitly obtained cryptocurrency into fiat money. This helps in tracing and potentially recovering the funds involved in ransomware attacks.
3. Offensive Measures Against Cybercriminals: The Department of Defense's "Defend Forward" strategy involves US Cyber Command actively targeting and

disrupting cybercriminals abroad. This proactive stance allows for the disruption of cybercriminal operations, utilizing the military's significant resources in addressing the ransomware problem.

These measures indicate that blockchain and cryptocurrency regulation play critical roles in both defending against ransomware attacks and disrupting the operations of cybercriminals. The focus on improving cybersecurity infrastructure, enhancing the visibility of cryptocurrency transactions, and proactively targeting cybercriminals illustrates a comprehensive strategy to mitigate the impact of ransomware.

Illustrating the intensifying crackdown on crypto exchanges enabling ransomware payments, Binance and its CEO pleaded guilty to federal charges in November 2023, resulting in a hefty $4 billion settlement. Investigations revealed that a key element of the indictment focused on hundreds of millions of dollars in illicit funds from ransomware attacks, dark web transactions, and various online scams flowing through Binance in a deliberate attempt to circumvent law enforcement detection (Department of Justice-1 2023).

6.3   Challenges and Considerations in Utilizing Web3 for Ransomware Defense

While Web3 technologies, especially blockchain, offer promising solutions against ransomware attacks, they are not without their challenges. This section delves into the potential pitfalls, limitations, and considerations that organizations and individuals should be aware of when leaning on Web3 as a defense mechanism against ransomware.

6.3.1   Scalability and Performance Concerns

The blockchain, decentralized storage, and smart contract solutions covered in this chapter offer enhanced security and resilience against ransomware attacks. However, real-world deployment of these technologies surfaces scalability and performance constraints that must be addressed.

Public blockchains provide immutable storage and transparency but are limited in transaction processing speed. Ethereum, the most common smart contract platform, handles only 15 transactions per second (Mart and Dempsey 2021)—woefully inadequate for enterprise data volumes. Committing large datasets directly on-chain would grind the network to a halt. Even processing access logs or threat alerts from across an organization could choke the blockchain.

Decentralized storage via platforms like IPFS allows securely dispersing data across nodes, but reconstructing files from fragmented pieces incurs latency, bandwidth limits, and consistency issues at scale. The network speed depends on node proximity and responsiveness. Parsing and reassembling terabytes of enterprise data requires high throughput exceeding most decentralized storage capabilities currently.

Smart contracts automate security responses like halting processes or denying unauthorized encryption. However, they are computationally restricted when running analytics algorithms or simulation models needed for identifying ransomware behavior. The execution costs quickly become prohibitive on public chains.

These performance barriers necessitate a hybrid architecture with the partitioning of data and processes between blockchain and conventional systems. Blockchain is well-suited for critical subsets of data where integrity is paramount, such as access logs, controls, and notifications. Storage, backup, threat detection, and remediation functions with high-volume data or computations can remain on legacy infrastructure.

Adopting this hybrid model entails upgrading legacy components for tight integration with blockchain security. Databases require hardened perimeter defenses like firewalls, network segmentation, rigorous access controls, and encryption to prevent central points of failure. Load balancers and caches can ease throughput constraints for decentralized storage reconstruction procedures. APIs streamline smart contract connectivity with off-chain analytics engines.

Ongoing research on blockchain scaling provides hope for overcoming current limitations. Sharding solutions like Ethereum 2.0 aim to partition nodes handling transactions, substantially increasing throughput as the network grows (Millman et al. 2022). Optimistic and zk-Rollups move bulk transaction processing off-chain with periodic settlement on-chain, reducing fees and latency (Pandey 2023). As these solutions mature, more enterprise data and workloads can shift to blockchain-centric security architectures.

Until then, the hybrid model balancing blockchain and legacy infrastructure provides the pragmatic path for leveraging decentralized security. The trick lies in hardening all aspects of the joint solution without overlooking the gaps ransomware can penetrate. Though a significant undertaking, the enhanced resilience and automation provided by blockchain and smart contracts offer vital supplementary defenses against the growing threat of ransomware.

6.3.2   Integration with Legacy Systems

While decentralized solutions like blockchain and IPFS offer security advantages, most enterprises today run on legacy infrastructure not inherently compatible with Web3 technologies. Integrating these disparate systems poses arduous technical and operational challenges.

Many legacy databases and data stores predate the inception of blockchain and lack native compatibility. They use centralized, client-server models allowing for high performance but low resilience. Blockchains operate on decentralized peer-to-peer topologies optimized for integrity over speed. Bridging these divergent architectures requires bespoke integration code and costly changes to legacy backend processes.

Legacy user interfaces are designed for client-server systems, not peer-to-peer decentralization. Modern single-page applications fetch data from APIs tied to centralized application servers and databases. Transitioning UIs to pull data from Web3 sources like IPFS or blockchain nodes requires re-architecting frontend code and connectivity. The overall user experience suffers from decentralized storage and slower reconstruction times.

On the operational side, legacy teams' skills center on maintaining high-availability systems through careful change control, access restrictions, and business continuity planning. Blockchain solutions shift control away from centralized administrators toward community consensus mechanisms unfamiliar to traditional operators. The paradigm shift can encounter organizational resistance and a lack of capable talent.

Cybersecurity policies and practices must also be revamped to address decentralized environments. Standards like PCI DSS have long-standing guidelines for firewalls, intrusion detection, and vulnerability scanning focused on perimeter-based threats (Sanchez 2021). Blockchain's transaction transparency and threat prevention through economic disincentives represents an entirely different model requiring new audit checklists, controls, and compliance procedures.

Further complicating integration is the rapid pace of innovation across Web3 solutions vying for enterprise adoption. As blockchain platforms like Ethereum, Cosmos, and Polygon evolve, integration code risks frequent breaking changes. Decentralized storage options like Filecoin and Storj take divergent architectural approaches. Each integration becomes its own one-off effort with significant maintenance costs as components rapidly change.

To navigate these turbulent waters, enterprises should take an iterative, project-based approach when embarking on ransomware resilience initiatives involving blockchain or decentralized storage. Focus on a narrowly defined use case like access controls or disaster recovery plan data integrity checks. Deliver tangible security value while minimizing the scope of integration touch points, allowing legacy systems to adjust gradually. Emphasize crypto-agility and modular architecture to swap emerging solutions as the market matures.

The road toward full integration of modern decentralized solutions with legacy environments remains long. But judicious, focused deployments to enhance ransomware protection provide a pathway for incremental adoption while strengthening enterprise defenses against an ever-expanding threat.

6.3.3 Economic and Operational Costs

Implementing emerging Web3 technologies as complementary ransomware defenses comes with substantial economic and operational costs beyond just the technical integration challenges. These expenses accumulate across various dimensions—infrastructure, development, organizational realignment, and compliance overhead.

At the infrastructure level, blockchain and decentralized storage solutions require extensive node deployment and management, especially for enterprise data volumes. Both on-premises and cloud-hosted options carry significant hardware and hosting costs surpassing those of traditional centralized infrastructure. Ongoing node maintenance tasks like security patching, performance monitoring, and disaster recovery further add to operational workloads.

These decentralized networks also incur transaction fees and smart contract execution costs, dubbed "gas" prices, on platforms like Ethereum (Garnett 2023). While negligible for minimal use cases, enterprise volumes on public chains result in gas fees accumulating to significant sums paid to validators and miners. Enterprise Ethereum alternatives like Hyperledger reduce expenses but still carry licensing and support costs (Gillis 2022).

Integrating and customizing these complex Web3 solutions requires specialized blockchain developers and engineers, currently among the most scarce and expensive technology skills. The learning curve is steep even for seasoned engineers used to client-server paradigms. Many consultancies and system integrators have responded with teams of highly paid blockchain developers. Even internal training or hiring such skills can set enterprises back six figures per engineer. Transitioning legacy interfaces and data pipelines from a centralized architecture to a decentralized model can significantly increase costs during the re-architecting process.

On the policy and process side, new frameworks must be developed to govern decentralized infrastructure and development lifecycles. Decentralized autonomous organizations introduce new entity structures with implications on governance, liability, and compliance. The highly fluid nature of the Web3 space also demands more nimble release processes compared to change-averse legacy environments.

Most profoundly, organizations must realign from long-held centralization tendencies. IT administrators give up significant control while business leaders adapt to decentralized transparency and censorship resistance. User experience and productivity metrics focused on speed and availability get deprioritized for resilience. This cultural shift is profound after decades of centralized systems, requiring exceptional leadership and change management.

Further inflating expenses are new compliance and audit demands introduced by decentralization, like validating chain integrity, controlling private keys, and documenting self-executing smart contract logic flows. External audits checklist blockchain environment controls and attest to their reliability for financial reporting. Failing these audits results in remediation costs or loss of license for regulated industries like finance.

In total, a multi-million dollar expenditure facing personnel constraints, cultural transitions and policy uncertainty represents a hard sell for making decentralized solutions the cornerstone of ransomware defense. But a focused strategy minimizing integration touch points while maximizing cyber resilience delivers functionality impossible in centralized-only environments. And as more digital assets move to blockchain and Web3 adoption advances, developing such skills, experience, and hardened defenses becomes a necessity not just for ransomware protection but business viability overall.

6.4 Conclusion

Ransomware represents a continually escalating threat landscape that demands proactive solutions to mitigate its impact. As cybercriminals grow more sophisticated in their tactics, all organizations remain vulnerable regardless of sector or size. Payment of ransom demands provides no guarantee of data recovery, making prevention and resilience top priorities.

This chapter has explored blockchain technology and its applications as a defense mechanism against ransomware. Blockchain's decentralized architecture offers vital advantages in integrity, transparency, and censorship resistance that thwart typical ransomware behavior. Core capabilities including immutable records, smart contracts, and decentralized data storage provide robust checks against unauthorized data encryption, modification, or denial of access.

However, blockchain and Web3 solutions have shortcomings around scalability, ease of integration, and implementation costs that necessitate tradeoffs. A pragmatic approach involves hybrid models judiciously applying blockchain security to critical subsets of data and systems, while traditional defenses fortify the rest of the infrastructure. As blockchain platforms mature to enterprise-grade capabilities, more workloads can transition to decentralized models.

Ultimately, battling ransomware's economic incentives requires coordinated deterrence from law enforcement, policymakers, and technology firms. But by progressively decentralizing systems, enhancing transparency and automating incident response through blockchain, organizations can significantly strengthen their resilience and reduce the crippling impact of attacks. Combined with workforce education, cyber insurance, and improved data governance, blockchain delivers vital tools in this constantly evolving fight against digital extortion.

References

Alspach, K. (2023, October 19). *Hackers Hit The IT Industry: 12 Companies Targeted In 2023*. CRN. Retrieved from https://www.crn.com/news/security/hackers-hit-the-it-industry-12-companies-targeted-in-2023

American University. (2023, February 23). *Combating Ransomware: One Year On*. Digital Commons @ American University Washington College of Law. Retrieved from https://digitalcommons.wcl.american.edu/cgi/viewcontent.cgi?article=1085&context=research

Blosil, J. (2022, October 24). *Ransomware cost: Measuring the true cost of a ransomware attack*. NetApp. Retrieved from https://www.netapp.com/blog/ransomware-cost

Department of Justice. (2021, July 15). *U.S. Government Launches First One-Stop Ransomware Resource at StopRansomware.gov*. Department of Justice. Retrieved from https://www.justice.gov/opa/pr/us-government-launches-first-one-stop-ransomware-resource-stopransomwaregov

Department of Justice-1. (2023, November 21). *Binance and CEO Plead Guilty to Federal Charges in $4B Resolution*. Department of Justice. Retrieved from https://www.justice.gov/opa/pr/binance-and-ceo-plead-guilty-federal-charges-4b-resolution

Franklin, R., & Tozzi, C. (2022, November 14). *RSA Encryption vs AES Encryption: What Are the Differences?* Precisely. Retrieved from https://www.precisely.com/blog/data-security/aes-vs-rsa-encryption-differences

Garnett, A. G. (2023, November 12). *What Is ETH Gas? Ethereum Fees & Gwei Explained*. Britannica. Retrieved from https://www.britannica.com/money/ethereum-gas-fees-eth

Gihon, S. (2023, July 4). *Ransomware Trends 2023, Q2 Report*. Cyberint. Retrieved from https://cyberint.com/blog/research/ransomware-trends-q2-2023-report/

Gillis, A. S. (2022). *What is Hyperledger? Everything You Need to Know*. TechTarget. Retrieved from https://www.techtarget.com/searchcio/definition/Hyperledger

Hussey, M. (2020, November 26). *What Is Filecoin and How Does It Work?* Decrypt. Retrieved from https://decrypt.co/49625/what-is-filecoin-and-how-does-it-work

Jacobi, J. L. (2023, July 27). *Storj review: Fast and affordable online storage for techies*. PC World. Retrieved from https://www.pcworld.com/article/1965026/storj-online-storage-review-25gb-free-storage-super-affordable-pay-plans-not-as-easy-as-it-could-be.html

Mart, J., & Dempsey, C. (2021, November 23). *Scaling Ethereum & crypto for a billion users*. Coinbase. Retrieved from https://www.coinbase.com/blog/scaling-ethereum-crypto-for-a-billion-users

McKay, D. (2022, February 10). *What Is the Interplanetary File System (IPFS) and How Do You Use It?* How-To Geek. Retrieved from https://www.howtogeek.com/784295/what-is-the-interplanetary-file-system-ipfs/

Millman, R., Kelly, L. J., & Graves, S. (2022). *What is Ethereum 2.0? Ethereum's Consensus Layer and Merge Explained*. Decrypt. Retrieved from https://decrypt.co/resources/what-is-ethereum-2-0

Osborne, C. (2023, June 1). *Global Ransomware Damage Costs Predicted To Exceed $265 Billion By 2031*. Cybercrime Magazine. Retrieved from https://cybersecurityventures.com/global-ransomware-damage-costs-predicted-to-reach-250-billion-usd-by-2031

Palmer, D. (2020, July 23). *Ransomware attack locked a football club's turnstiles*. ZDNet. Retrieved from https://www.zdnet.com/article/ransomware-attack-locked-a-football-clubs-turnstiles-almost-leading-to-cancelled-match/

Pandey, H. (2023, April 26). *A Deep Dive into Optimistic and Zk Rollups for Blockchain Scalability*. Medium. Retrieved from https://medium.com/coinmonks/a-deep-dive-into-optimistic-and-zk-rollups-for-blockchain-scalability-dc9904c0d123

Perlroth, N. (2018, March 27). *A Cyberattack Hobbles Atlanta, and Security Experts Shudder (Published 2018)*. The New York Times. Retrieved from https://www.nytimes.com/2018/03/27/us/cyberattack-atlanta-ransomware.html

Poggi, N. (2021). *Types of Encryption: Symmetric or Asymmetric? RSA or AES?* | Prey Blog. Prey Project. Retrieved from https://preyproject.com/blog/types-of-encryption-symmetric-or-asymmetric-rsa-or-aes

Sanchez, A. (2021, August 27). *The 12 PCI DSS Compliance Requirements*. Alert Logic. Retrieved from https://www.alertlogic.com/blog/the-12-pci-dss-compliance-requirements/

Singh, A., & Ali, M. A. (2022). *Blockchain: Tool for Controlling Ransomware through Pre-Encryption and Post-Encryption Behavior*. 2022 Fifth International Conference on Computational Intelligence and Communication Technologies (CCICT). https://ieeexplore.ieee.org/document/9913615

SpyCloud. (2023). *The Hidden Costs of Ransomware Attacks*. SpyCloud. Retrieved from https://spycloud.com/blog/the-hidden-costs-of-ransomware-attacks/

Ton, J. (2022, April 8). *Ransomware Damage: Are You Forgetting About Your Reputation?* Forbes. Retrieved from https://www.forbes.com/sites/forbestechcouncil/2022/04/08/ransomware-damage-are-you-forgetting-about-your-reputation/

Waldman, A. (2023, January 6). *10 of the biggest ransomware attacks of 2022*. TechTarget. Retrieved from https://www.techtarget.com/searchsecurity/news/252528956/10-of-the-biggest-ransomware-attacks-of-2022?Offer=abt_pubpro_AI-Insider

Waterson, J. (2022, December 21). *Guardian hit by serious IT incident believed to be ransomware attack*. The Guardian. Retrieved from https://www.theguardian.com/media/2022/dec/21/guardian-hit-by-serious-it-incident-believed-to-be-ransomware-attack

Jerry Huang has worked as a technical and security staff at several prominent technology companies, gaining experience in areas like security, AI/ML, and large-scale infrastructure. At Metabase, an open-source business intelligence platform, he contributed features such as private key management and authentication solutions. As a Software Engineer at Glean, a Generative AI search startup, Jerry was one of three engineers responsible for large-scale GCP infrastructure powering text summarization, autocomplete, and search for over 100,000 enterprise users. Previously at TikTok, Jerry worked to design and build custom RPCs to model access control policies. And at Roblox, he was a Machine Learning/Software Engineering Intern focused on real-time text generation models. He gathered and cleaned a large multilingual corpus that significantly boosted model robustness. Jerry has also conducted extensive security and biometrics research as a Research Assistant at Georgia Tech's Institute for Information Security and Privacy. This resulted in a thesis on privacy-preserving biometric authentication. His academic background includes a BS/MS in Computer Science from Georgia Tech and he is currently pursuing an MS in Applied Mathematics at the University of Chicago.

Ken Huang is the author and chief editor of eight books on Generative Artificial Intelligence and Web3, published respectively by international publishers including Springer, Cambridge University Press, John Wiley, and China Machine Press. He currently serves as the CEO of the AI and Web3 consulting and education company DistributedApps.AI, based in the United States. Additionally, he holds multiple roles including the expert member of the Blockchain Committee of the Chinese Institute of Electronics, the Co-Chair of the AI Organization Responsibility Working Group at Cloud Security Alliance, and Chair of the Blockchain Security Working Group at the Cloud Security Alliance, GCR. He is also a core contributor to the Generative AI Working Group at the NIST and a core author of the OWASP Top 10 for LLM Applications.

Ken Huang has been invited to provide Speaking or Consulting services at institutions including the University of California, Berkeley, Stanford University, Peking University, Tsinghua University, Shanghai Jiao Tong University, China Pacific Insurance, and the World Bank in the past.

- Moreover, he has given keynote speeches at international conferences, such as:
- The Davos World Economic Forum 2020 Blockchain Conference
- Consensus 2018 in New York
- The American ACM AI & Blockchain Decentralized Annual Conference 2019
- IEEE Technology and Engineering Management Society Annual Meeting 2019
- Silicon Valley World Digital Currency Forum
- Sino-US Blockchain Summit in Silicon Valley

He has also been awarded the "Blockchain 60" Figure Award by the National University Artificial Intelligence and Big Data Innovation Alliance Blockchain Special Committee in China in 2021.

Chapter 7
Web3 and Supply Chain Risks

Jerry Huang ⓘ, Ken Huang ⓘ, and Sean Heide

Abstract This chapter examines the vulnerabilities inherent in the Web3 ecosystem, particularly focusing on the risks arising from the reliance on third-party open-source libraries and frameworks, as well as external service providers. Through detailed case studies, the chapter highlights how these vulnerabilities can be exploited, causing significant disruptions in the Web3 ecosystems. The chapter provides valuable insights into the potential risks and offers strategies for mitigating these challenges, aiming to foster more secure software development and risk management practices within the Web3 ecosystem.

Web3 applications rely heavily on third-party open-source libraries and frameworks, which promotes rapid development and innovation. However, this also introduces potential vulnerabilities that attackers can exploit. The supply chain risks for Web3 applications are multifaceted, stemming not just from vulnerable software but also from service providers—whether honest mistakes or unexpected downtimes from cloud providers hosting Dapp UIs or Oracle data feeders network issues. This chapter explores the realm of supply chain risks in Web3, offering insights through case studies and proposing countermeasures.

While we discuss risks, it is important to note that Web3 technologies like blockchain can positively enhance supply chain transparency and provenance tracking in areas like agriculture, food, art, and more. However, this chapter focuses narrowly on two facets of risk: software vulnerabilities introduced through open-source reuse in Web3 applications, and risks that arise from dependence on external service providers underpinning Web3 architectures. By delineating these risks with case

J. Huang
The University of Chicago, Chicago, IL, USA

K. Huang (✉)
DistributedApps LLC, Fairfax, Virginia, USA
e-mail: ken@distributedapps.ai

S. Heide
Cloud Security Alliance, Seattle, WA, USA

© The Author(s), under exclusive license to Springer Nature
Switzerland AG 2024
K. Huang et al. (eds.), *Web3 Applications Security and New Security Landscape*,
Future of Business and Finance, https://doi.org/10.1007/978-3-031-58002-4_7

studies, we aim to raise awareness and inspire more secure software development and risk mitigation practices across the growing Web3 ecosystem.

7.1   Understanding Supply Chain Risks in Web3

The supply chain in the context of Web3 extends beyond physical goods and services—it embodies the flow of data, code, and digital services. This section offers an overview of what constitutes supply chain risks in the world of Web3, setting the foundation for deeper exploration into vulnerabilities and potential defenses.

As shown in Fig. 7.1, we mainly discuss two kinds of risks: supply chain code risks and supply chain service risks.

7.1.1   Web3 Supply Chain Code Risks

The tapestry of Web3 applications is rich and diverse, and much of its rapid expansion can be attributed to the adoption of open-source libraries and frameworks. These resources, available to developers worldwide, have been a beacon of collaborative development. They have democratized the creation process, enabling even those without extensive resources to craft intricate applications (Linux Foundation, 2023). However, as the old adage goes, every rose has its thorns. The very aspects that make open-source attractive also open the door to a myriad of vulnerabilities.

Open-source libraries act as the building blocks for many Web3 applications. These pre-built modules, smart contracts, open protocol implementations, and functions allow developers to sidestep the time-consuming process of crafting every component from scratch. Instead, they can focus on integrating these blocks into a cohesive application, dramatically accelerating the development timeline. This speed, combined with the cost savings associated with bypassing proprietary software licenses, makes open source an almost irresistible choice for many.

Beyond the practicality, there is an ideological allure to open source. It embodies the spirit of collective effort, where developers, regardless of their geographical

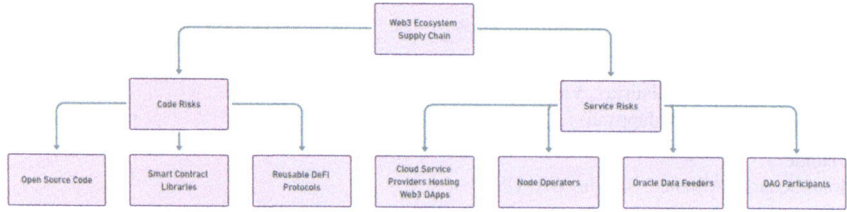

Fig. 7.1 Web3 ecosystem supply chain risks

location or affiliations, contribute to a shared goal. This pooling of global talent often results in code that is not just functional but continually refined and optimized. A single library can benefit from the expertise of countless professionals, each bringing their unique perspective to the table.

However, the landscape is not entirely idyllic. The transparency of open source, while being one of its greatest strengths, is also a potential weakness. When multiple Web3 applications lean on the same open-source library, they inadvertently introduce a common point of vulnerability (see 7.2.1). A flaw or oversight in that library does not just compromise one application; it jeopardizes every application that has integrated that particular piece of code. This domino effect means that an attacker, armed with the knowledge of a single vulnerability, has the potential to wreak havoc on a broad scale.

The decentralized ethos of open-source development also presents oversight challenges. Unlike proprietary software, which is often developed under the watchful eyes of dedicated teams within structured environments, open-source projects can sometimes lack stringent vetting processes. This environment can be a double-edged sword. While it fosters innovation and inclusivity, it can also allow vulnerabilities to slip through the cracks. Worse still, it can provide an avenue for malicious actors to deliberately introduce flaws or backdoors, camouflaged amidst legitimate contributions. As indicated in a research by Vitalik Buterin, there are five common types of backdoor problems, such as Arbitrary Transfer, and he found that 189 confirmed backdoor threats were detected from real-world smart contracts (Buterin, 2022).

Even when vulnerabilities are detected and patches are developed, open-source projects might suffer from delays in their dissemination. The decentralized nature of the ecosystem means that there is not always a streamlined mechanism to notify users of these updates. This delay creates a perilous window during which applications remain vulnerable, giving attackers ample opportunity to exploit known flaws. One such famous example was the DAO attack while the vulnerability was disclosed and discussed long before the actual attack happened (Siegel, 2016).

Compounding the challenge is the intricate web of dependencies that many Web3 applications possess. It is not uncommon for an application to rely on multiple open-source libraries, each with its own set of dependencies. This intricate hierarchy makes it daunting to track vulnerabilities, ensure compatibility, and keep every component updated.

7.1.1.1 Smart Contract Libraries: The Achilles' Heel of Web3

Smart contracts are self-executing contracts with the terms of the agreement directly written into code. They are fundamental to the functionality of decentralized applications (dApps), especially in the DeFi sector. While smart contract libraries provide reusable code that accelerates development, they also introduce critical risks.

One of the primary concerns is the immutable nature of smart contracts once deployed on the blockchain. If a smart contract contains a vulnerability, it cannot be easily rectified. This immutability, while a feature in terms of security and trust, becomes a significant liability in the face of code flaws. The DAO hack of 2016 is a prominent example, where a vulnerability in a smart contract led to the theft of millions of dollars worth of Ethereum (Siegel, 2016).

Furthermore, smart contracts are often composed of layered complexities, with one contract interacting with multiple others. This interconnectedness means that a vulnerability in one contract can have cascading effects on the entire system. The challenge is compounded by the lack of standardized practices for writing and testing smart contracts, leading to inconsistencies in quality and security across the ecosystem.

7.1.1.2 Reusable DeFi Protocols: Innovation Amidst Vulnerability

DeFi, or decentralized finance, represents a significant shift from traditional financial systems, offering more accessible, efficient, and transparent financial services. Reusable DeFi protocols, also called "Money Legos," form the backbone of this sector, providing standardized processes for activities like lending, borrowing, and trading.

However, the rapid growth of DeFi, coupled with the experimental nature of many of its protocols, introduces substantial risks. Many DeFi protocols are built on complex smart contracts that are not fully tested in real-world scenarios. This untested nature, combined with the high financial stakes involved, makes DeFi protocols prime targets for exploits.

Interoperability, while a key advantage in DeFi, also increases risk exposure. Protocols often interact with multiple other applications and smart contracts, creating a web of dependencies. A vulnerability in one protocol can quickly propagate through this web, affecting multiple platforms and users.

The composability of DeFi protocols, where one protocol can easily integrate or build upon another, further amplifies these risks. While this promotes innovation and rapid development, it also means that security vulnerabilities can be inadvertently introduced and perpetuated across the ecosystem (Chap. 1 of this book has more detail on DeFi security risks).

7.1.1.3 Addressing the Risks

Mitigating the risks associated with Web3 supply chain code requires a comprehensive and multifaceted approach, encompassing:

1. Enhanced Security Practices for Open-Source Projects: This includes regular security audits, robust maintenance and patching protocols, and the establishment of community-driven governance models to oversee project development.
2. Rigorous Testing and Standardization of Smart Contracts: Developing standardized frameworks and best practices for smart contract development is crucial.

This includes extensive testing, both in simulated environments and through bug bounty programs, to identify and rectify vulnerabilities before deployment.

3. Risk Assessment and Management in DeFi Protocols: Implementing thorough risk assessment frameworks to evaluate the security and stability of DeFi protocols before integration. Continuous monitoring and updating of these protocols are essential to respond to emerging threats and vulnerabilities.

4. Education and Community Engagement: Raising awareness about the inherent risks in Web3 development among stakeholders, including developers, users, and investors, is vital. This includes educating them about best practices in code development, usage, and investment in Web3 platforms.

5. Regulatory and Legal Frameworks: Collaborating with regulatory bodies to develop legal frameworks that ensure accountability and consumer protection.

7.1.2 Service Provider Vulnerabilities

In the intricate mosaic of Web3's ecosystem, service providers play a pivotal role. They offer the infrastructure that powers the decentralized web, facilitating hosting, data storage, transaction validations, and myriad other essential functions. While the software and its open-source underpinnings form the bedrock of Web3, these service providers act as the scaffolding, enabling the software to function seamlessly and efficiently in real-world scenarios. However, as with every facet of this dynamic ecosystem, relying on third-party service providers comes with its own set of vulnerabilities and challenges.

Web3's decentralized promise is built upon a network of nodes, often maintained by different service providers, each with its own operational protocols, security measures, and technical competencies. This diversity, while fostering resilience against centralized failures, also introduces multiple potential points of vulnerability. If one provider in this interconnected web falters, either due to technical glitches or security breaches, the ripples can be felt throughout the network.

One of the most benign yet disruptive vulnerabilities stems from honest mistakes. No system is infallible, and even the most diligent service provider can encounter unexpected downtimes, data losses, or misconfigurations. In a traditional centralized setup, the repercussions of such errors might be limited to that singular entity. But in the Web3 paradigm, an honest mistake by one provider can cascade, causing disruptions that affect a vast array of applications and end users. For instance, if a hosting provider faces an unexpected outage, every application relying on that provider could become inaccessible, undermining the reliability of the decentralized web.

Deliberate malfeasance presents an even graver concern. While the decentralized nature of Web3 offers robustness against single points of failure, it is not entirely immune to targeted attacks. A malicious actor, intent on compromising the system, might focus on service providers as potential entry points. By breaching the defenses of a single provider, the attacker could potentially gain access to a treasure trove of data or even exert influence over parts of the network. This scenario is not just hypothetical; there have been instances where service providers in the blockchain space have been targeted, leading to significant data leaks and financial losses.

Furthermore, the economic models underlying some service providers could also introduce vulnerabilities. Many providers operate on slim margins, continually seeking to optimize costs. This cost cutting could, at times, come at the expense of robust security measures or the adoption of the latest protective technologies.

In navigating the promise and potential of Web3, understanding the landscape is crucial. While the software and code form the core, the role of service providers cannot be understated. As organizations and developers embrace the decentralized future, vigilance is paramount. It is essential to assess the reliability, security protocols, and track records of service providers. By ensuring that every link in the chain, from software to service, is robust and secure, the true potential of Web3 can be realized, unhindered by vulnerabilities and unmarred by breaches.

7.1.2.1 Cloud Service Providers Hosting Web3 dApps

Cloud service providers are integral to the Web3 infrastructure, offering the computational power and storage necessary for hosting decentralized applications. While they bring scalability and efficiency, their centralized nature poses several risks:

1. Single Point of Failure: Relying heavily on centralized cloud services can negate the decentralized nature of Web3. If a cloud provider faces downtime or security breaches, it could incapacitate all dApps reliant on its services.
2. Data Privacy and Security: Cloud providers store vast amounts of data, making them lucrative targets for cyber-attacks. Any compromise in their security systems could lead to significant data breaches, affecting all hosted dApps.
3. Vendor Lock-in: Dependency on a specific cloud provider can lead to vendor lock-in, where switching providers becomes difficult and costly. This can limit the flexibility and agility of dApp developers.

7.1.2.2 Node Operators

Node operators maintain the blockchain's integrity by validating transactions and blocks. They are crucial for the decentralized consensus mechanisms in Web3. However, they also present risks:

1. Centralization of Nodes: If a significant number of nodes are controlled by a few operators, it could lead to centralization, making the blockchain vulnerable to manipulation and reducing its security.
2. Malicious Actors: Rogue node operators can potentially disrupt the network by refusing to validate transactions or by validating fraudulent transactions.

3. Technical Failures: Nodes operated with insufficient security or technical capabilities can become points of failure, potentially slowing down or halting network operations.

7.1.2.3 Oracle Data Feeders

Oracles provide external data to smart contracts on the blockchain, enabling them to execute based on real-world events and information. However, the reliability of smart contracts is directly tied to the integrity of the Oracle data:

1. Data Manipulation: If the data fed by an oracle is inaccurate or tampered with, it can trigger incorrect contract executions, leading to losses or unintended outcomes.
2. Single Source Dependency: Relying on a single oracle for data makes the smart contract vulnerable to outages or malfunctions from that oracle.
3. Privacy Concerns: Oracles that handle sensitive data need robust privacy and security measures to prevent data breaches or misuse.

7.1.2.4 DAO Participants

DAOs are organizational structures run through rules encoded in smart contracts. While they promote decentralized governance, there are inherent risks:

1. Governance Vulnerabilities: If a small group of participants gains a majority of the governance tokens, they can exert disproportionate control over the DAO, potentially leading to decisions that are not in the best interest of all stakeholders.
2. Smart Contract Flaws: DAOs are only as strong as their underlying smart contracts. Any vulnerabilities in these contracts can be exploited, jeopardizing the entire DAO.
3. Legal and Regulatory Uncertainty: DAOs operate in a novel legal terrain, and regulatory uncertainties can pose risks concerning compliance, liability, and enforcement.

7.1.2.5 Mitigation Strategies

To address these risks, several mitigation strategies can be employed:

1. Diversification and Redundancy: Using multiple cloud providers, node operators, and oracles can reduce dependency on any single entity, enhancing the resilience of the system.
2. Robust Security Measures: Implementing state-of-the-art security protocols for data protection, regular security audits, and employing decentralized oracles can mitigate the risks of data manipulation and breaches.

3. Decentralized Governance Models: Designing DAOs with checks and balances, transparent decision-making processes, and broad participation can help mitigate governance risks.
4. Compliance with Regulations: Staying informed and compliant with emerging regulations can help navigate legal uncertainties and enhance trust and legitimacy.
5. Community Engagement and Education: Fostering a knowledgeable and vigilant community can help in early detection and response to potential threats or vulnerabilities.

While Web3 technologies offer groundbreaking opportunities for decentralized operations, they also introduce specific service risks in the supply chain. Understanding, anticipating, and mitigating these risks is crucial for the secure and efficient functioning of the Web3 ecosystem.

7.2 Case Studies of Web3 Supply Chain Breaches

Real-world incidents offer invaluable lessons. This section delves into notable case studies where Web3 applications faced supply chain breaches. By understanding these incidents, we can glean insights into common patterns, attacker motivations, and potential defenses.

7.2.1 Vulnerable Library and Multiple Web3 Apps

In the ever-evolving sphere of Web3, where innovation and rapid deployment are paramount, the allure of open-source libraries is undeniable. They serve as shortcuts, allowing developers to integrate functionalities without starting from scratch. However, this reliance on a shared resource led to a cascading vulnerability that affected a swathe of Web3 applications, offering a poignant lesson on the double-edged nature of open-source dependencies.

In December 2023, a vulnerability in an open-source library was reported. The vulnerability is common across the Web3 space and impacted the security of multiple NFT collections, including Coinbase (Toulas, 2023). This event is particularly notable due to its potential impact on a diverse array of smart contracts, which are fundamental to the operation and trustworthiness of various Web3 applications, including notable NFT collections and pre-built smart contracts.

The gravity of this situation is underscored by the fact that, although the vulnerability had not been exploited at the time of its discovery, it presented a significant threat. The potential for unauthorized access to sensitive data and control over the affected applications could have had far-reaching consequences, jeopardizing the integrity and reliability of the decentralized web. The affected contracts, which included key formats such as AirdropERC20, ERC721, and ERC1155, are integral to a wide range of Web3 functionalities.

The release of a patch and the urgent call for smart contract owners to implement necessary mitigation steps demonstrate a responsible and swift approach to managing such critical security risks. This incident serves as a powerful reminder of the inherent interconnected risks within the Web3 ecosystem. It emphasizes the necessity for continuous vigilance, rigorous vetting of software components, and a collaborative approach to security across all stakeholders in the decentralized space.

Moreover, this incident raises critical questions about the reliance on third-party libraries in the development of smart contracts and other Web3 applications. While the use of open-source libraries is a common practice, offering significant benefits such as reduced development time and access to a wealth of community-tested features, this scenario illuminates the flip side of such dependencies. It highlights the paramount importance of conducting thorough security evaluations and ongoing monitoring of these libraries to ensure the safety and security of the Web3 ecosystem.

The incident brings to the forefront the concept of shared responsibility in maintaining the security of the decentralized web. This encompasses not only the developers and entities directly involved in Web3 development but also extends to the wider community, including users and stakeholders. It underscores the need for a unified approach to security, where best practices are shared, vulnerabilities are promptly reported, and patches are swiftly implemented.

The vulnerability also serves as a stark reminder of the potential risks associated with the rapid pace of innovation in the Web3 space. The balance between advancing technological frontiers and ensuring the security and reliability of these innovations is delicate. It necessitates a comprehensive approach to security that includes not only technological solutions but also a strong focus on education and awareness among developers and users alike.

In light of this, the role of platforms like Thirdweb becomes increasingly significant. As a hub for Web3 development, such platforms bear the responsibility of not only providing tools and services but also ensuring that these offerings are secure and reliable. The incident demonstrates the critical role that these platforms play in safeguarding the ecosystem against potential threats and vulnerabilities.

Furthermore, this situation highlights the ongoing challenge of securing decentralized systems. Unlike traditional centralized architectures, where security measures can be more uniformly applied and managed, the decentralized nature of Web3 presents unique challenges. The distributed and often open-source nature of these systems requires a more nuanced approach to security, involving a combination of technical safeguards, community engagement, and continuous monitoring and updating of security protocols.

Another notable example is the Ankr protocol incident on the BNB chain, which occurred on December 2, 2022 (Science, 2023). This event was a stark reminder of the potential vulnerabilities in decentralized finance (DeFi) platforms. The exploit involved the unauthorized minting of an astronomical 10 trillion aBNBc tokens due to a compromise of a governance key. This breach resulted in a significant drain from the DEX pool, culminating in a loss of around $5 million. The investigation into this incident uncovered that the breach was orchestrated by former members of the Ankr team, who executed a supply chain attack by injecting malicious code into the software packages. This code was designed to compromise Ankr's private keys during routine project updates. This situation was exacerbated when the flooded market with aBNBc tokens led to another exploit, impacting the Helio Protocol and resulting in an additional loss of approximately $15.5 million.

These cases underscore the critical need for robust security measures in the Web3 space, particularly around code dependencies and governance mechanisms. They serve as a reminder of the ongoing and evolving challenges in ensuring the integrity and security of decentralized systems and the importance of vigilance and proactive measures to mitigate such risks.

7.2.2 Service Provider Downtime and Cascading Failures

The vast expanse of Web3 is built upon a delicate interplay of various components working in harmony. These components, ranging from open-source libraries to dedicated service providers, form the intricate lattice that upholds the decentralized world. However, every so often, a misstep or unforeseen glitch in one component can trigger a cascade of disruptions, revealing the inherent fragility of this interconnected ecosystem.

7.2.2.1 AMS Outage and Its Impact on Web3

One such incident, rooted not in malice but in unforeseen technical challenges, highlighted the perils of overreliance on singular entities in the Web3 domain.

An Amazon Web Service (AWS) outage in December, 2021, forced the decentralized exchange dYdX to halt operations, raising concerns over the reliance on centralized services by DeFi protocols (Newar, 2021). AWS is one of the most widely used cloud services in the world, and many decentralized infrastructure projects rely on it for their operations. AWS offers servers, storage, networking, remote computing, email, mobile development, and security for websites.

dYdX acknowledged that its reliance on a centralized web service like AWS is problematic and pledged to improve the true decentralization of its operations, although it did not specify how. The outage affected not only dYdX but also centralized crypto exchanges like Binance.US and Coinbase. dYdX is the eleventh biggest DeFi app on the Ethereum network, with about $1.5 billion in daily trading volume. As a decentralized exchange (DEX), it does not have a know-your-customer (KYC) protocol and settles all transactions via smart contracts. The AWS outage prevented dYdX from accessing key elements of its back end, and the team was unable to determine if it could cancel all orders. This incident highlights the challenges faced by decentralized platforms in ensuring true decentralization while still relying on centralized services.

The ramifications were not just technical. Financial losses mounted as transactions were stalled. Trust, a cornerstone of the Web3 ethos, was eroded as users questioned the wisdom of centralizing critical functionalities with singular service providers, no matter how reputable.

In the aftermath, the Web3 community grappled with pressing questions. How could they ensure redundancy in their operations to prevent similar incidents? Was it prudent to distribute dependencies across multiple service providers, even if it meant higher operational complexities? And crucially, how could they strike a balance between efficiency and resilience?

The incident served as a clarion call for diversification and redundancy in the Web3 world. It underscored the importance of contingency planning and the need to anticipate and mitigate single points of failure. While the cloud service provider swiftly rectified the glitch and restored operations, the lessons from the incident lingered. It became a testament to the fact that in the interconnected world of Web3, resilience and adaptability are not just virtues but necessities.

7.2.2.2 Attacks on Blockchain Nodes

Web3 applications rely on blockchain nodes to validate transactions. As such, blockchain nodes are the most important supply chain service nodes for Web3 applications. These nodes are not bullet proof. If it is Proof of work nodes and if the hash power is not enough, the whole blockchain network can get 51% attack. Some examples of blockchains that have suffered 51% attacks include Bitcoin SV, Verge, Ethereum Classic, Hanacoin, Vertcoin, Expanse, Litecoin Cash, and Bitcoin Gold. These attacks have demonstrated the potential vulnerability of certain blockchains to the control of more than 50% of the network's hashing power, leading to the ability to alter the blockchain, execute double spending, or prevent transactions from being confirmed. The 51% attack is a significant concern for blockchain networks, especially those with lower hashing power, as it can undermine the security and trustworthiness of the blockchain. While the attack is theoretically limited in the amount of disruption it can cause, it remains a serious threat, and robust security measures are essential to prevent such attacks (Lovejoy, 2022).

7.2.2.3 Wallet Supply Chain Vulnerability

Wallet is the foundational component for all Web3 applications, and as such wallet service provider provides part of supply chain service for Web3. It is important to make sure that this type of service has maximum security possible.

On Ledger, a popular crypto hardware and software wallet provider, announced that it had been the target of a supply chain attack, which compromised the code behind a crypto protocol used by multiple Web3 applications and services. The hackers replaced a genuine version of Ledger Connect Kit, a library used by decentralized apps (dApps) to connect to the Ledger wallet service, with a malicious version. Ledger has since pushed out a genuine version to replace the malicious file and urged users not to interact with any dApps until further notice.

Ledger has sold 6 million units of its hardware wallet and has 1.5 million users on its software equivalent, Ledger Live. This incident highlights the importance of security in the Web3 ecosystem and the potential risks associated with various service providers including wallet providers (Franceschi-Bicchierai, 2023).

As another example, the Atomic Wallet, a popular mobile and desktop crypto wallet, was hacked, resulting in the theft of over \$35 million in various cryptocurrencies. The attack, which occurred in early June 2023, led to reports of compromised wallets and the loss of funds from users' accounts. The largest single victim reportedly lost almost \$8 million, and the five biggest individual losses accounted for \$17 million. The investigation into the incident revealed that the North Korean hacking group Lazarus was linked to the attack, and the group was using various methods to launder the stolen funds. The developers of Atomic Wallet took down their download server and are working with third-party security companies to investigate the incident and block the stolen funds from being sold on exchanges. The US Department of the Treasury has sanctioned the Sinbad cryptocurrency mixing service for its alleged use by North Korean hackers who have performed large-scale crypto heists, including the Atomic Wallet hack. The hackers responsible for the Atomic Wallet attack have used cross-chain liquidity protocol THORChain to conceal their stolen funds. The attack on Atomic Wallet is part of a series of high-profile hacks attributed to North Korean state hackers, who have been linked to significant cryptocurrency thefts since 2017, totaling an estimated \$3 billion (Abrams, 2023).

7.2.2.4 Oracle Nodes Supply Chain Risk

Web3 applications such as DeFi and NFT applications rely heavily on the Oracle network to provide a price feed to their smart contracts in order for them to work. Any vulnerability of the Oracle network is a significant supply chain risk to Web3 applications.

Some examples of DeFi issues caused by Oracle price failures include:

1. Mango Markets Oracle Hack: Approximately \$112 million in digital assets was lost due to an Oracle hack, highlighting the significant impact of Oracle failures on DeFi platforms (QuillAudits, 2023).
2. Compound Lending Platform Liquidation: In 2020, the manipulation of the Dai stablecoin price led to the liquidation of around \$89 million on the Compound lending platform, demonstrating the substantial financial impact of Oracle price manipulation (Chipolina, 2020).
3. Chainlink's Dominance as a Risk Dependency: While Chainlink's dominance represents a risk dependency, it is not a single point of failure. However, data concentration at the DeFi protocol level poses risks associated with Oracle failures (S&P Global, 2023).

These examples underscore the potential for substantial financial losses and disruptions within the DeFi landscape due to Oracle price failures, emphasizing the critical importance of robust and reliable Oracle solutions in safeguarding the integrity of DeFi platforms.

7.3   Web3 SupplyChain Risks Mitigation

While the risks are manifold, so are the defenses. This section delves into strategies and mechanisms that can be employed to mitigate supply chain risks in the Web3 ecosystem, ensuring resilience and security. Figure 7.2 depicts high-level strategies of mitigation and we will explore in detail in each subsection.

7.3.1   Auditing and Continuous Monitoring

In the ever-evolving landscape of Web3, maintaining a robust defensive posture is not just about reacting to threats, but proactively identifying and mitigating them. The cornerstone of this proactive approach lies in regular auditing and continuous monitoring, ensuring that vulnerabilities, whether latent or emerging, are promptly identified and addressed.

Auditing, especially of open-source dependencies, is a crucial first step. Given the decentralized nature of open-source development, where contributions come from diverse quarters, it is imperative to regularly review and vet the codebase. A comprehensive audit scrutinizes every line of code, ensuring that it adheres to best practices, is free from known vulnerabilities, and has not been tampered with for malicious intent. Engaging external cybersecurity firms for these audits can offer an unbiased perspective, bringing to light vulnerabilities that might be overlooked internally due to familiarity or oversight.

While audits provide a snapshot of the codebase's health at a given moment, the dynamic nature of Web3 means that threats can emerge at any time. This is where continuous monitoring steps in, acting as the ecosystem's ever-watchful sentinel. Monitoring tools, equipped with real-time scanning capabilities, track changes, updates, and modifications to the codebase. By cross referencing these changes against databases of known vulnerabilities or anomalous patterns, these tools can flag potential threats as they arise.

But the realm of continuous monitoring is not confined to code alone. With the growing complexities of Web3 applications and their interactions with various service providers, monitoring must encompass the entire ecosystem. This includes tracking the operational health of service providers, ensuring uptime, and being

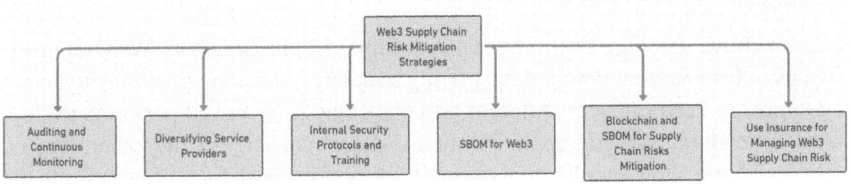

Fig. 7.2 Web3 supply chain risks mitigation strategies

alert to any signs of compromise. Furthermore, transaction monitoring can detect irregular patterns, such as sudden surges in fund transfers or unusual smart contract activations, indicating potential breaches.

Implementing best practices in both auditing and monitoring necessitates a multi-pronged approach. Utilizing tools that harness machine learning and artificial intelligence can enhance detection capabilities, enabling the identification of novel threats or sophisticated attack patterns. Additionally, fostering collaboration within the Web3 community, where organizations share insights, threat intelligence, and defense strategies, can amplify collective defenses.

As the Web3 universe continues its expansive trajectory, the challenges of safeguarding its supply chain will undoubtedly grow. However, with a rigorous regimen of auditing and continuous monitoring, bolstered by the latest tools and a collaborative spirit, the ecosystem can stay a step ahead of potential threats. Ensuring that the decentralized future is not just innovative, but also secure and resilient.

7.3.2   Diversifying Service Providers

In the intricate network of the Web3 ecosystem, reliance on service providers is a given. These entities offer the essential infrastructural backbone, be it through hosting, data storage, or even transaction validations. However, an age-old adage in finance, "Don't put all your eggs in one basket," rings particularly true in this context. Concentrating too much dependency on a single service provider, no matter how reputable or reliable, can be a recipe for vulnerability.

Centralizing resources or functionalities under a single provider's umbrella can introduce a significant risk factor: the single point of failure. Should this provider face a technical glitch, security breach, or even operational challenges like unexpected downtime, the consequences can be catastrophic for all Web3 applications tethered to it. The domino effect of such an incident can reverberate across the ecosystem, leading to widespread disruptions, loss of data, and even financial implications.

Diversifying service providers is a strategic move to distribute and thereby mitigate this risk. By splitting dependencies across multiple providers, the potential impact of any single provider's failure is contained. If one provider faces an issue, backup services from other providers can ensure continued functionality, minimizing disruptions.

Beyond mere risk mitigation, diversifying providers also offers several other advantages:

1. Competitive Pricing: Engaging with multiple providers allows Web3 applications to benefit from competitive pricing structures, ensuring cost-effectiveness.
2. Optimized Performance: Different providers may offer varied performance levels depending on geographic locations, server capacities, and other factors. Diversifying allows applications to optimize performance by routing tasks to the best-suited provider.

3. Enhanced Security: Leveraging multiple providers can be a boon for security. With different providers often employing varied security protocols and measures, the chances of a universal breach diminish.
4. Negotiation Leverage: A diversified approach ensures that no single provider becomes indispensable, granting applications better negotiation leverage when discussing terms, conditions, and pricing.

However, while diversification brings with it numerous benefits, it is not without challenges. Managing relationships with multiple providers can be complex, requiring more administrative oversight. Integration can also be more intricate, given the potential variations in APIs, protocols, and standards across different providers.

To navigate these challenges, it is essential to have a clear diversification strategy. This includes thorough vetting of providers, understanding their strengths and potential weaknesses, and ensuring seamless interoperability between them. Regular performance and security audits across all providers can ensure that the standards remain consistent.

By embracing a multi-provider approach, Web3 applications can not only enhance their resilience but also optimize performance, security, and cost effectiveness. In the ever-evolving digital frontier, where challenges are aplenty, diversifying service providers emerges as a prudent strategy to navigate potential pitfalls and harness the promise of decentralization to its fullest.

7.3.3 Internal Security Protocols and Training

An organization's internal controls and policies serve as critical safeguards against both unintentional errors and intentional misconduct from within. As the old adage states, a chain is only as strong as its weakest link; in the Web3 context, this underscores the importance of shoring up internal defenses.

Robust internal security controls form the foundation on which a culture of vigilance and accountability can be built. Well-designed controls provide a framework guiding all members of an organization. These include access restrictions delineating permitted data and systems based on roles, along with multifactor authentication adding extra verification layers to ensure sensitive information and crucial operations remain protected.

However, even the most sturdy controls can fall short without ongoing, comprehensive training to complement them. The digital environment constantly evolves, with new threats rising and old ones shifting form. Keeping fully up-to-date on these changes and ensuring all staff, from junior personnel to senior leadership, can recognize and mitigate them is imperative. Standardized training programs, workshops, and simulated attacks can help guarantee everyone from new hires to veteran developers have mastered the latest threats and effective countermeasures.

Beyond the technicalities, training also plays a pivotal role in fostering a culture of security awareness. When every member feels a sense of ownership and responsibility toward safeguarding the organization's digital assets, the collective defenses are naturally heightened. This sense of collective vigilance becomes especially crucial in countering insider threats. With regular training, individuals can not only recognize potential external threats but also spot anomalous behaviors or actions from peers, nipping potential threats in the bud.

Additionally, an open and transparent organizational culture can act as a deterrent against insider threats. When grievances can be aired, and there is a genuine avenue for redressal, feelings of discontent—which can sometimes be the precursors to insider breaches—are less likely to fester.

In essence, while the external realm of firewalls, encryptions, and blockchain defenses is vital, the internal world of protocols, training, and culture holds equal, if not more, significance. As Web3 continues its march toward becoming the digital standard, ensuring that the very individuals who propel this revolution are well equipped, well trained, and instilled with a sense of shared responsibility is the key to a secure and prosperous decentralized future.

7.3.4   SBOM for Web3

A Software Bill of Materials (SBOM) is a comprehensive inventory of the components that are included in a piece of software. Essentially, it is a detailed list that provides visibility into the software's composition, specifically detailing all of its elements. This inventory typically includes information about direct and indirect dependencies, such as libraries, packages, and modules used in the software (Kerner, 2022).

The components listed in an SBOM can be open-source or proprietary, free or paid, and widely available or restricted. The level of detail in an SBOM can vary but generally includes the following:

1. Component Names: The names of all the software components or libraries used.
2. Version Information: Specific versions or version ranges of these components.
3. Licenses: Licensing information for each component, which is crucial for legal compliance.
4. Source References: Information about where each component is sourced from, such as a repository URL.
5. Dependencies: Information on how components are related or dependent on each other.
6. Hashes or Checksums: These provide a way to verify the integrity of the components.

The primary purpose of an SBOM is to ensure transparency and improve the security of software. It helps software developers, users, and security professionals to:

– Identify Vulnerabilities: Quickly determine if a software product is dependent on a component with known vulnerabilities.

- Compliance and Licensing: Ensure compliance with licenses and avoid potential legal issues related to software dependencies.
- Security and Risk Management: Assess the security risks associated with each component, especially important in environments where security is critical.
- Software Supply Chain Management: Track and manage the various components sourced from different suppliers, especially in complex software ecosystems.

The concept of an SBOM has gained significant attention in the context of software supply chain security, especially following high-profile software supply chain attacks. Governmental organizations, like the United States' National Institute of Standards and Technology (NIST), have recognized the importance of SBOMs in enhancing cybersecurity resilience (NIST, 2022). The development of SBOMs is also being encouraged and standardized across industries to foster greater transparency and security in software development and deployment.

7.3.4.1   SBOM in the Context of Web3

Web3 represents a shift from centralized to decentralized network architectures, primarily using blockchain technology. This decentralization brings unique challenges and opportunities in terms of security and transparency. In Web3, software components often include smart contracts, decentralized applications (dApps), and various blockchain protocols, each with their own dependencies and potential vulnerabilities.

An SBOM can help Web3 projects in identifying vulnerabilities, managing licenses, and ensuring the integrity of the software supply chain.

7.3.4.2   Relevance of SBOM to Web3 Challenges

1. Security and Vulnerability Management: The decentralized nature of Web3 applications makes them attractive targets for cyber-attacks. An SBOM provides visibility into the software components, helping developers and security professionals identify and remediate vulnerabilities swiftly.
2. Compliance and Licensing: Web3 projects often utilize a mix of open-source and proprietary software. An SBOM assists in managing compliance with software licenses, a critical aspect given the legal complexities in decentralized environments.
3. Transparency and Trust: Trust is a cornerstone of Web3 applications. An SBOM enhances trust by providing clear visibility into the software's makeup, crucial for users who rely on the integrity of decentralized applications.
4. Facilitating Audits: Blockchain and smart contract audits are vital for ensuring the security of Web3 applications. An SBOM serves as a key tool for auditors to understand the software's composition, making the audit process more efficient and thorough.

7.3.4.3 Integration with NIST SSDF

The National Institute of Standards and Technology's Secure Software Development Framework (NIST SSDF) provides guidelines for secure software development. Integrating SBOMs into Web3 aligns with several key practices recommended by the NIST SSDF:

- Prepare the Organization (PO): This involves ensuring that the organization's people, processes, and technology are prepared for secure software development. An SBOM aids in this preparation by offering a clear picture of the software components in use.
- Protect the Software (PS): This focuses on protecting all components of the software from tampering and unauthorized access. SBOMs enhance this protection by identifying components that need to be secured.
- Produce Well-Secured Software (PW): This practice emphasizes the importance of producing secure software by design. SBOMs contribute to this by enabling developers to identify and mitigate potential security issues in third-party components.
- Respond to Vulnerabilities (RV): This involves responding effectively to vulnerabilities in software. SBOMs are essential tools in identifying affected components and facilitating a swift response.

7.3.4.4 Alignment with Supply Chain Levels for Software Artifacts

Supply Chain Levels for Software Artifacts (SLSA) is a framework for ensuring the integrity of software artifacts throughout the software supply chain (Aqua Security, 2022). The use of SBOMs in Web3 aligns with SLSA's goals in several ways:

- Source Integrity: By listing the origins of software components, SBOMs help ensure that the source code comes from trusted entities.
- Built-In Controls: SBOMs support the SLSA's emphasis on automated controls that reduce the likelihood of tampering with software components.
- Provenance: SLSA stresses the importance of provenance or the origin and evolution of the software. SBOMs provide detailed provenance information, essential for tracing the history of each component.
- Dependency Tracking: SBOMs facilitate the tracking of dependencies, a key aspect of SLSA, ensuring that dependencies are secure and up-to-date.

7.3.4.5 Challenges and Considerations

While SBOMs offer numerous benefits, their implementation in Web3 poses specific challenges:

1. Dynamic Nature of Web3 Projects: Web3 projects often evolve rapidly, with frequent updates and changes. Maintaining an accurate and up-to-date SBOM can be challenging in this fast-paced environment.
2. Complex Dependencies: The dependencies in Web3 applications can be complex, especially with the integration of various blockchain protocols and smart contracts. Creating comprehensive SBOMs that capture this complexity is a nontrivial task.
3. Privacy Concerns: In some cases, revealing detailed information about software components through SBOMs might raise privacy concerns, particularly in decentralized applications where anonymity is valued.
4. Standardization: The lack of standardization in the format and content of SBOMs can lead to inconsistencies and difficulties in their use across different Web3 projects.

7.3.4.6 Future Directions

As Web3 continues to evolve, the role of SBOMs in ensuring the security and integrity of these new types of applications will likely become more prominent. Collaborative efforts between developers, security experts, and regulatory bodies will be essential in addressing the challenges and maximizing the benefits of SBOMs in this domain.

– Developing Standards: There is a need for industry-wide standards for SBOMs in Web3 to ensure consistency and interoperability.
– Automated Tools: The development of automated tools to generate and update SBOMs in real time could greatly enhance their efficacy in the dynamic Web3 environment.
– Education and Awareness: Raising awareness about the importance of SBOMs among Web3 developers and stakeholders is crucial for their widespread adoption.
– Integration with Other Security Practices: Combining SBOMs with other security practices and frameworks, such as SLSA and NIST SSDF, can create a more robust security posture for Web3 applications.

As we can see, SBOMs play a crucial role in enhancing the security and integrity of Web3 applications. By providing transparency, aiding in compliance, and facilitating vulnerability management, SBOMs contribute significantly to the trustworthiness and reliability of decentralized applications. The integration of SBOMs with frameworks like NIST SSDF (NIST-1, 2021) and SLSA further underscores their importance in a comprehensive approach to software security. However, the unique challenges posed by the Web3 ecosystem necessitate continued efforts in standardization, tool development, and community engagement to fully realize the potential of SBOMs in this emerging field.

7.3.5 Blockchain and SBOM for Supply Chain Risks Mitigation

In the context of SBOM, blockchain can be employed to record, store, and verify SBOM data, enhancing the integrity and trustworthiness of the information.

7.3.5.1 Enhancing SBOM with Blockchain: Key Advantages

1. Immutable Record Keeping: Once an SBOM is recorded on a blockchain, it cannot be altered or tampered with, ensuring the integrity of the software component data.
2. Transparency Across the Supply Chain: Blockchain's distributed nature allows all stakeholders in the software supply chain to access and verify SBOM data, fostering transparency and trust.
3. Real-Time Verification and Updates: Blockchain technology facilitates real-time updates and verification of SBOMs, essential in the dynamic landscape of software development where components are frequently updated or replaced.

7.3.5.2 Implementation Strategy

To effectively use blockchain for enhancing SBOM in supply chain code security, several steps, and considerations are involved:

1. Integration of SBOM with Blockchain Platforms: Developing a system where SBOM data can be seamlessly recorded and retrieved from a blockchain platform.
2. Stakeholder Collaboration: Ensuring all stakeholders, from developers to end users, understand and can interact with the blockchain-enabled SBOM system.
3. Compliance with Regulations: Aligning the blockchain-SBOM system with legal and regulatory requirements, especially concerning software security and data privacy.

7.3.5.3 Challenges and Solutions

While the potential benefits are significant, several challenges need addressing:

1. Scalability: Blockchain platforms must handle large volumes of SBOM data efficiently.
2. Interoperability: Ensuring the blockchain-SBOM system works seamlessly with various software development tools and platforms.
3. User Adoption: Encouraging the adoption of this new system across diverse stakeholders in the software supply chain.

7.3.5.4   The Future: Blockchain-Enhanced SBOM in Supply Chain Code Security

The future of using blockchain technology for SBOM in supply chain code security looks promising. Innovations and advancements in blockchain technology could further enhance the capabilities of SBOMs:

1. Smart Contracts for Automated Compliance: Implementing smart contracts on the blockchain to automatically enforce compliance with software security standards.
2. AI-Driven Analytics: Integrating AI to analyze SBOM data on the blockchain for predictive security insights.
3. Cross-Industry Standardization: Developing standardized protocols for blockchain-SBOM integration applicable across various industries.

The integration of blockchain technology with SBOMs presents a groundbreaking approach to securing software supply chains. By ensuring the integrity, transparency, and traceability of software components, this integration addresses critical challenges in software supply chain security. As the technology evolves, blockchain-enhanced SBOMs could become a standard practice, significantly bolstering the security and reliability of software systems in our increasingly digital world.

7.3.6   Use Insurance for Managing Web3 Supply Chain Risk

The integration of insurance and risk-sharing protocols in the Web3 ecosystem, particularly for supply chain code and supply chain providers, offers a reasonable risk management approach to deal with the inherent risks in this evolving digital landscape. Insurance and risk-sharing protocols tailored to Web3 can provide essential safety nets and risk mitigation strategies.

7.3.6.1   Insurance Protocols for Web3 Supply Chain Risk

The following are a few areas insurance protocols can help to manage Web3 supply chain risks:

Smart Contract Coverage: Insurance protocols can offer coverage against vulnerabilities and bugs in smart contracts, which are critical in Web3 supply chains. This coverage can safeguard against financial losses due to contract failures or exploits. Decentralized insurance protocols can offer coverage against vulnerabilities and bugs in smart contracts, safeguarding against financial losses due to contract failures or exploits. Smart contracts are self-validating and self-executing protocols that can be used in insurance for automation of various operations such as policy issuance, claim processing, and fraud detection. In the context of supply chains, smart contracts can help automate and streamline processes, reducing the risk of errors and increasing efficiency. Decentralized insurance platforms, such as Nexus Mutual (Strack, 2023) and InsurAce Protocol (InsurAce, 2023), provide coverage for events like smart contract exploits. While smart contracts offer benefits, they can also be prone to errors, making insurance coverage important to mitigate risks. The use of decentralized insurance and smart contracts has been proven to be an effective approach in mitigating the risks associated with DeFi activities and smart contract exploits.

Cybersecurity Insurance: Given the digital nature of Web3, cybersecurity risks are prominent. Insurance protocols can protect against data breaches, hacking incidents, and other cyber threats that could compromise the supply chain code.

Operational Risk Insurance: This covers risks associated with the operational aspects of Web3 technologies, including downtime due to Cloud provider's downtime or failures in decentralized applications (dApps) that are crucial in the supply chain or bugs in wallets.

7.3.6.2 Risk-Sharing Protocols for Web3 Supply Chain Providers

Web3 supply chains can leverage decentralized finance protocols to implement innovative risk management and risk-sharing models such as decentralized risk pools, tokenized risk exposures, and automated claims processing.

1. Decentralized Risk Pools: Risk-sharing protocols in Web3 can create decentralized pools where supply chain providers collectively contribute to and share risks. This approach distributes the impact of individual risks, reducing the burden on any single participant.
2. Tokenized Risk Management: These protocols can allow the tokenization of supply chain risks, enabling providers to trade or hedge their risk exposures in a decentralized marketplace. This system promotes a more efficient and transparent way of managing risks.
3. Automated Claims and Settlements: Leveraging smart contracts, these protocols can automate the claims process, ensuring timely and fair settlements without the need for intermediaries. This automation increases efficiency and reduces administrative overhead.

7.3.6.3 Future Prospects

The future of insurance and risk-sharing in Web3 supply chains looks promising, with potential advancements including integration of AI to provide predictive analytics for risk assessment and management in supply chain operations. Additional advancements could involve developing protocols that are compatible across different blockchain platforms to ensure broader applicability and flexibility, as well as offering more tailored insurance products and risk-sharing mechanisms that cater to the specific needs of diverse Web3 supply chain providers. Enhanced customization and flexibility of decentralized insurance protocols will help drive adoption among supply chain entities with varied risk profiles and operational models. Meanwhile, leveraging AI and pursuing cross-chain interoperability will maximize efficiency and utility across Web3 insurance and risk transfer platforms. Overall, the innovation potential is immense when it comes to decentralized finance advancing risk mitigation for next generation, blockchain-enabled supply chains.

7.4 Conclusion

Web3 represents a profound shift toward decentralization, promising to transform aspects of society from finance to supply chains. However, its reliance on open-source software and third-party service providers inevitably introduces risks that could undermine trust and adoption. As elucidated through case studies, vulnerabilities have manifested as exploitable software flaws, unexpected system failures, and malicious attacks, leading to technical outages and substantial financial losses.

Mitigating these multifaceted risks necessitates a multi-pronged approach spanning across the Web3 ecosystem. More rigorous auditing, diversification of dependencies, internal security protocols, SBOMs, blockchain integrations, and risk transfer mechanisms can all play pivotal roles. However, beyond technical defenses, fostering security awareness and best practices among developers and users is equally vital. Additionally, further research into novel decentralized architectures, transparency frameworks like SLSA, and specialized insurance protocols tailored for Web3 could unlock innovation in risk management.

By recognizing these challenges early in its evolution, the Web3 community has an invaluable opportunity. Through collective vigilance, responsibly balancing rapid innovation with security, this next-generation Internet can manifest its fullest potential. The blockchain-powered future offers immense promise but securing it demands proactive commitment from all stakeholders. If undertaken in earnest, Web3 could pioneer "trust minimization" in the true sense, where decentralization and cryptography minimize the need for third-party trust—creating an ecosystem resilient against compromise and worthy of widespread confidence. With patient nurturing, its sprouts of disruption can mature into an equitable and secure digital society.

References

Abrams, L. (2023, June 4). *Atomic Wallet hacks lead to over $35 million in crypto stolen.* Bleeping Computer. Retrieved from https://www.bleepingcomputer.com/news/security/atomic-wallet-hacks-lead-to-over-35-million-in-crypto-stolen/

Aqua Security. (2022, October 19). *What Is SLSA and How to Use it for Supply Chain Security.* Aqua Security. Retrieved from https://www.aquasec.com/cloud-native-academy/supply-chain-security/slsa/

Buterin, V. (2022, September 20). *Research Summary: Pied-Piper: Revealing the Backdoor Threats in Ethereum ERC Token Contracts.* Smart Contract Research Forum. Retrieved from https://www.smartcontractresearch.org/t/research-summary-pied-piper-revealing-the-backdoor-threats-in-ethereum-erc-token-contracts/2083

Chipolina, S. (2020, November 26). *Oracle Exploit Sees $89 Million Liquidated on Compound.* Decrypt. Retrieved from https://decrypt.co/49657/oracle-exploit-sees-100-million-liquidated-on-compound

Franceschi-Bicchierai, L. (2023, December 14). *Supply chain attack targeting Ledger crypto wallet leaves users hacked.* TechCrunch. Retrieved from https://techcrunch.com/2023/12/14/supply-chain-attack-targeting-ledger-crypto-wallet-leaves-users-hacked/

InsurAce. (2023, April 20). *InsurAce & LI.FI Unveil Bridge Cover, the Ultimate Safety Net for Cross-Chain DeFi Transactions.* Yahoo Finance. Retrieved from https://finance.yahoo.com/news/insurace-li-fi-unveil-bridge-130000749.html

Kerner, S. M. (2022). *What is software bill of materials (SBOM)?* TechTarget. Retrieved from https://www.techtarget.com/whatis/definition/software-bill-of-materials-SBOM

Linux Foundation. (2023, February 28). *Why the future of Web3 needs open source, sustainable blockchains.* Linux Foundation. Retrieved from https://www.linuxfoundation.org/blog/why-the-future-of-web3-needs-open-source-sustainable-blockchains

Lovejoy, J. (2022). *51% Attacks — MIT Digital Currency Initiative.* MIT Digital Currency Initiative. Retrieved from https://dci.mit.edu/51-attacks

Newar, B. (2021, December 10). *AWS outage hits dYdX, raising concerns over its decentralization.* Cointelegraph. Retrieved from https://cointelegraph.com/news/aws-outage-hits-dydx-raising-concerns-over-decentralization

NIST. (2022, May 3). *Software security in supply chains: Software bill of materials (SBOM).* National Institute of Standards and Technology. Retrieved from https://www.nist.gov/itl/executive-order-14028-improving-nations-cybersecurity/software-security-supply-chains-software-1

NIST-1. (2021, February 25). *Secure Software Development Framework | CSRC.* NIST Computer Security Resource Center. Retrieved from https://csrc.nist.gov/projects/ssdf

QuillAudits. (2023, September 20). *Oracle Failures in DeFi: Causes, Consequences, and Solutions.* QuillAudits Blog. Retrieved from https://blog.quillaudits.com/2023/09/20/oracle-failures-in-defi-causes-consequences-and-solutions/

Science, M. (2023, January 6). *Hack Track: Analysis of Ankr Exploit.* Merklescience. Retrieved from https://blog.merklescience.com/general/hack-track-analysis-of-ankr-exploit?hs_amp=true

Siegel, D. (2016, June 25). *The DAO attack: Understanding what happened—CoinDesk.* CoinDesk. Retrieved from https://www.coindesk.com/learn/understanding-the-dao-attack/

S&P Global. (2023, November 17). *Utility at a cost: Assessing the risks of blockchain oracles.* S&P Global. Retrieved from https://www.spglobal.com/en/research-insights/featured/special-editorial/utility-at-a-cost-assessing-the-risks-of-blockchain-oracles

Strack, B. (2023, February 17). *Nexus mutual pays out over $5M to FTX, BlockFi users.* Blockworks. Retrieved from https://blockworks.co/news/payouts-to-ftx-blockfi-users

Toulas, B. (2023, December 5). *Multiple NFT collections at risk by flaw in open-source library.* Bleeping Computer. Retrieved from https://www.bleepingcomputer.com/news/security/multiple-nft-collections-at-risk-by-flaw-in-open-source-library/

Jerry Huang has worked as a technical and security staff at several prominent technology companies, gaining experience in areas like security, AI/ML, and large-scale infrastructure. At Metabase, an open-source business intelligence platform, he contributed features such as private key management and authentication solutions. As a Software Engineer at Glean, a Generative AI search startup, Jerry was one of three engineers responsible for large-scale GCP infrastructure powering text summarization, autocomplete, and search for over 100,000 enterprise users. Previously at TikTok, Jerry worked to design and build custom RPCs to model access control policies. And at Roblox, he was a Machine Learning/Software Engineering Intern focused on real-time text generation models. He gathered and cleaned a large multilingual corpus that significantly boosted model robustness. Jerry has also conducted extensive security and biometrics research as a Research Assistant at Georgia Tech's Institute for Information Security & Privacy. This resulted in a thesis on privacy-preserving biometric authentication. His academic background includes a BS/MS in Computer Science from Georgia Tech and he is currently pursuing an MS in Applied Mathematics at the University of Chicago. phone: 571-268-6923; Address: Jerry Huang, The University of Chicago, 5801 S EllisAve, Chicago, IL 60637, USA

ORCID:0009-0005-6224-4785

Email: jerryh@uchicago.edu

Ken Huang is the author and chief editor of eight books on Generative Artificial Intelligence and Web3, published respectively by international publishers including Springer, Cambridge University Press, John Wiley, and China Machine Press. He currently serves as the CEO of the AI and Web3 consulting and education company DistributedApps.AI, based in the United States. Additionally, he holds multiple roles including the expert member of the Blockchain Committee of the Chinese Institute of Electronics, the Co-Chair of the AI Organization Responsibility Working Group at Cloud Security Alliance, and Chair of the Blockchain Security Working Group at the Cloud Security Alliance, GCR. He is also a core contributor to the Generative AI Working Group at the NIST and a core author of the OWASP Top 10 for LLM Applications.

Ken Huang has been invited to provide Speaking or Consulting services at institutions including the University of California, Berkeley, Stanford University, Peking University, Tsinghua University, Shanghai Jiao Tong University, China Pacific Insurance, and the World Bank in the past.

Moreover, he has given keynote speeches at international conferences, such as:

The Davos World Economic Forum 2020 Blockchain Conference

Consensus 2018 in New York

The American ACM AI & Blockchain Decentralized Annual Conference 2019

IEEE Technology and Engineering Management Society Annual Meeting 2019

Silicon Valley World Digital Currency Forum

Sino-US Blockchain Summit in Silicon Valley

He has also been awarded the "Blockchain 60" Figure Award by the National University Artificial Intelligence and Big Data Innovation Alliance Blockchain Special Committee in China in 2021.

Sean Heide , MSc has spent over a decade working on cyber security and cloud computing research and development. As a published author, public speaker, and researcher, Sean has led initiatives on major research releases and security frameworks that have been utilized by Fortune 500 companies. Sean spent 8 years in the US Naval Intelligence field before transitioning fulltime to cyber security. Sean holds a master's degree in information security management with a specialization in cyber security.

Email: sheide@cloudsecurityalliance.org

Chapter 8
Web3 and AI Security

Jerry Huang (ID), **Ken Huang** (ID), **Krystal Jackson, Luyao Zhang** (ID), **and Jennifer Toren**

Abstract This chapter examines the intersection of Web3 and AI in security, focusing on how they can enhance each other. It highlights AI's potential for dynamic threat detection and adaptive security in Web3 environments. However, AI also introduces risks like generative output manipulation, overreliance, and ethical concerns. To address these, the chapter proposes governance frameworks encompassing model verification, multi-layered defense strategies, and responsible AI development leveraging Web3's decentralized nature. Perspectives are provided on constraints for AI advancement, integrating AI in critical infrastructure, data-centric machine learning pipelines with DLT, model integrity via blockchain, and mitigating AI existential risk through Web3.
This chapter examines the synergy between Web3 technologies and AI in enhancing mutual security.

The objective of this chapter is not to cover the broad scope of AI Security. For this, readers can refer to our Springer book on Generative AI Security (Huang et al., 2024). This chapter also does not aim to address the broad scope of Web3 security. Readers are encouraged to refer to our book "A Comprehensive Guide for Web3 Security: From Technology, Economic and Legal Aspects" published by Springer in 2023 for a thorough overview.

J. Huang
The University of Chicago, Chicago, IL, USA

K. Huang (✉)
DistributedApps LLC, Fairfax, Virginia, USA
e-mail: ken@distributedapps.ai

K. Jackson
University of California, Berkeley, Berkeley, USA

L. Zhang
Duke Kunshan University, Suzhou, Jiangsu, P.R. China

J. Toren
Cloud Security Alliance, Seattle, WA, USA

© The Author(s), under exclusive license to Springer Nature
Switzerland AG 2024
K. Huang et al. (eds.), *Web3 Applications Security and New Security Landscape*,
Future of Business and Finance, https://doi.org/10.1007/978-3-031-58002-4_8

153

Rather, this chapter focuses specifically on the intersection of Web3 and AI in the security domain. It explores in detail how Web3 technologies can enhance AI security as well as how AI innovations can bolster security in Web3 environments. The discussions delve into the synergies between these two realms, analyzing how they complement each other to create more robust and adaptive security solutions. By highlighting the interplay of Web3 and AI in strengthening security, this chapter provides targeted insights beyond a general overview of either domain independently. The objective is to focus on their symbiosis in security contexts rather than their breadth as stand-alone subjects.

8.1 The Promise of AI in Web3 Security

AI's reasoning ability exhibited via ChatGPT and the potential of proactive detection capabilities have positioned it as an excellent tool in Web3 security. This section explores the myriad ways AI enhances security in the Web3 ecosystem, particularly through leveraging blockchains as the infrastructure that supplies abundant open-source data, vital for the applications of machine learning (Zhang, 2023a).

8.1.1 Dynamic Web3 Threat Detection by AI

In the digital realm, the game of cat and mouse between security experts and cyber attackers is relentless. As soon as one vulnerability is patched or one type of attack is mitigated, a new threat emerges from the shadows. This ever-evolving landscape has long posed a challenge for traditional security measures which, despite their effectiveness, are often reactive in nature. They require knowledge of a threat to protect against it, creating a window of vulnerability between the emergence of a new threat, its identification, and the subsequent response. Enter AI—a tool that promises not just to level the playing field, but to tip the scales in favor of defense.

AI's core strength in security stems from its ability to learn and adapt rapidly. Instead of relying solely on predefined threat databases or static algorithms, a well-designed AI-driven security solutions can continuously assimilate new data, refine their algorithms, and enhance their reasoning and threat detection capabilities. This continuous learning mechanism, rooted in machine learning, allows AI tools to not just recognize known threats but also to identify patterns, anomalies, or behaviors indicative of previously unseen cyberattacks.

For instance, consider the realm of malware detection. Traditional antivirus solutions largely depend on signature-based detection, where they match files against a database of known malware signatures. While effective against known threats, this approach struggles against zero-day attacks or newly developed malware. An AI-powered solution, on the other hand, does not just look for known signatures. It evaluates the behavior of files, analyzing them for patterns consistent with malicious activities. If a file starts acting similarly to known malware, even if its signature is unfamiliar, the AI system can flag or quarantine it.

Beyond malware, AI's dynamic detection capabilities extend to other areas of cyber threat, such as phishing attacks. Sophisticated phishing attempts nowadays can easily bypass traditional filters, especially those that employ machine learning tactics to craft convincing fake emails. However, AI can be trained to recognize subtle clues in language patterns, sender–receiver relationships, and even the timing of the emails, offering a more nuanced and effective defense against phishing.

Moreover, AI's capabilities are not limited to threat detection. Once a potential threat is identified, AI systems can also aid in swift and efficient threat response. Through predictive analytics, AI can forecast the potential trajectory of an attack, enabling preemptive measures. For example, if an AI system detects an unusual outbound data flow from a particular server, it can not only flag this activity but also predict which other servers or data banks might be targeted next, allowing for proactive defense measures.

The use of vast datasets for training AI algorithms adds another layer of advantage. By processing and learning from vast troves of cyber threat intelligence, AI models can glean insights from global threat landscapes. This means that an attack or a novel threat vector identified in one part of the world can enhance the AI's defense capabilities universally, offering protection even to those who have not encountered that specific threat yet.

Generative AI can play a crucial role in enhancing security in the Web3 ecosystem. Web3 platforms are inherently different from traditional web platforms due to their decentralized nature, smart contract-based operations, and reliance on blockchain technology. This distinct architecture exposes them to unique security vulnerabilities, such as smart contract bugs, blockchain-specific attacks (like 51% attacks), and decentralized application vulnerabilities.

As shown in Fig. 8.1, Generative AI can be employed to dynamically detect and prevent these threats in several ways:

1. Anomaly Detection: Generative AI models can be trained to understand normal blockchain and DApp operations. These models can then identify anomalies in transaction patterns, smart contract executions, or network activities, flagging potential security threats like fraudulent transactions or unusual contract interactions.
2. Smart Contract Auditing: By training on a vast corpus of smart contract code, generative AI can analyze new or existing contracts for vulnerabilities. It can

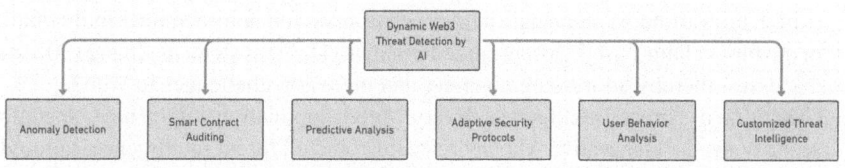

Fig. 8.1 Dynamic Web3 threat detection by AI

suggest optimizations or highlight potential security flaws, such as reentrancy attacks or overflow/underflow issues, before they are exploited.

3. Predictive Analysis: Generative AI can forecast potential attack vectors by analyzing trends in blockchain security incidents. For instance, a generative AI model might analyze historical data of security breaches involving smart contracts on the Ethereum blockchain. By doing so, it could uncover common patterns in the code that were exploited, such as reentrancy attacks or gas limit issues. The model can then generate predictions about what kinds of vulnerabilities might be targeted next, based on emerging trends in attack methodologies or changes in the blockchain ecosystem. Such predictive analysis is invaluable for blockchain developers and security teams. It enables them to anticipate and pre-emptively address potential vulnerabilities in their systems. For example, if the AI predicts a rise in attacks exploiting a specific smart contract function, developers can proactively review and strengthen the security of any contracts that use this function. Similarly, if the AI forecasts new types of phishing attacks targeting cryptocurrency users, wallet providers can implement additional security measures or user education programs to mitigate these risks.

4. Adaptive Security Protocols: In a Web3 environment, blockchain protocols and applications might need to evolve rapidly. Generative AI can assist in creating adaptive security measures that evolve based on the changing landscape of threats and vulnerabilities in the Web3 space.

    For example, imagine a newly developed blockchain protocol designed to support a decentralized social media platform in the Web3 environment. This protocol must ensure the integrity and privacy of user data while remaining open to public participation and interaction. Generative AI can be instrumental in creating adaptive security measures for this protocol.

5. User Behavior Analysis and Augmentation: Understanding user behavior is crucial in the development and use of decentralized applications. Generative AI can analyze user interactions with DApps to identify potentially malicious activities, like phishing or exploitation of DApp vulnerabilities. For example, a generative AI system can be used to enhance security on a decentralized exchange (DEX) by analyzing user interactions. It monitors transaction patterns and wallet activities, identifying anomalies that could indicate malicious behavior, such as phishing attempts or exploitation of smart contract vulnerabilities. The AI distinguishes between normal and suspicious activities by learning from ongoing user interactions. Once it detects potential threats, the DEX can take immediate action, like restricting transactions or alerting users, thereby improving its overall security against cyber threats in real time. Moreover, Generative AI can enhance user experiences to boost financial literacy, especially for non-technical individuals in the blockchain-based Decentralized Finance (DeFi) system, as Zhang (2023b) discusses, thereby addressing diversity and inclusion challenges in Web3.

6. Customized Threat Intelligence: Every Web3 application might have specific security needs. Generative AI can create customized threat intelligence, tailored to the particular requirements and threat landscape of individual DApps. For example, in a Web3 gaming DApp, generative AI can analyze the unique interac-

tions and transactions specific to the gaming platform, identifying potential security risks such as in-game asset theft or manipulation of game mechanics. By focusing on the specific context and user behavior patterns within this gaming DApp, the AI provides targeted intelligence on threats that are most relevant, enabling the implementation of precise and effective security measures.

For implementing such a system, your development approach would involve integrating these AI capabilities with your existing Web3 stack. This could mean creating AI models that interact with your blockchain data, smart contracts, and user interfaces developed using React Native and Ether.js. It would also involve rigorous testing and continuous training of the AI models using the latest data on Web3 threats and vulnerabilities.

It's important to note that while AI can significantly enhance security, it should be used as a part of a comprehensive security strategy that includes traditional security measures, such as code audits, rigorous testing, and user education. As Web3 platforms continue to evolve, staying ahead in security will be a continuous process, requiring both advanced technologies like AI and sound security practices.

8.1.2 AI-Enabled Adaptive Security Posture for Web3

Leveraging AI to enable an adaptive security posture in Web3 environments can significantly enhance the protection of decentralized applications and platforms. AI's capabilities in data analysis, pattern recognition, and automated response make it an ideal tool for addressing the dynamic and complex security landscape of Web3. The decentralized nature of Web3, characterized by blockchain technology, smart contracts, and DeFi applications, presents unique challenges that AI can help to mitigate.

In addition to what we have discussed in Sect. 8.1.1, we see a few other areas that AI can help Web3 to adapt to new security landscapes (as shown in Fig. 8.2).

1. Enhanced Transaction Monitoring with Machine Learning: AI, especially machine learning models, can be trained to understand normal transaction patterns within a blockchain network. These models can then detect deviations from these patterns, flagging potentially fraudulent or malicious activities. For example, 51% attack on proof of work blockchain or malicious transaction patterns on a DeFi application.
2. Dynamic Risk Assessment Tools: AI can dynamically assess the risk levels of various transactions or smart contracts based on historical data and evolving patterns. By continuously learning from new transactions and outcomes, AI models can provide up-to-date risk assessments, helping users and systems make informed decisions about engaging with certain contracts or entities. For example, consider "SafeDeFi," a fictitious decentralized finance (DeFi) platform using AI to assess the risk of lending agreements and investment contracts. This AI system, trained on historical transaction data and market trends, dynamically

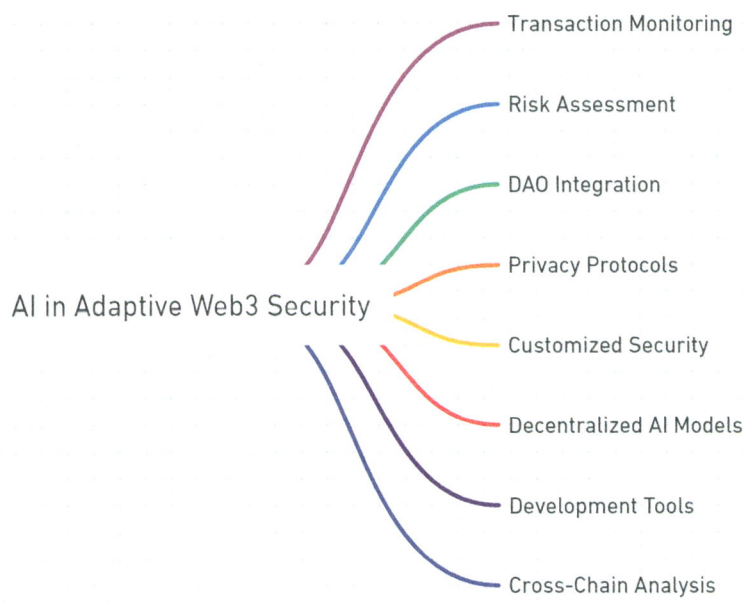

Fig. 8.2 AI in adaptive Web3 security

evaluates the risk of each loan application. It analyzes factors like the borrower's blockchain transaction history, collateral asset volatility, and current market conditions. Based on this analysis, the AI generates a risk score that influences the loan's interest rate and collateral requirements. Continuously updated with new data, this AI model ensures up-to-date risk assessments, helping SafeDeFi manage risk effectively and protect its users' investments.

3. Integration with Decentralized Autonomous Organizations (DAOs): AI can be integrated into the governance mechanisms of DAOs, aiding in decision-making processes. For example, AI can analyze proposals, predict their potential impacts based on historical data, and provide recommendations to the members of the DAO.

4. AI in Enhancing Privacy Protocols: AI can contribute to the development of advanced privacy-preserving technologies in Web3. For instance, AI algorithms and GPU chips can optimize the performance of zero-knowledge proofs (Burger, 2022), a cryptographic method that allows one party to prove to another that a statement is true, without revealing any information about the statement itself.

5. Customized User Experience and Security: AI can tailor the security experience based on individual user behavior and preferences. By understanding each user's interaction patterns, AI can adjust security measures accordingly, enhancing both user experience and security.

6. Decentralized AI Models for Security: Leveraging decentralized AI models, which are distributed and operate on a blockchain, can enhance the security and transparency of AI operations in Web3. These models can provide verifiable and transparent AI services without central points of failure.

7. AI-Driven Development and Testing Tools: AI can assist in the development of Web3 applications by providing advanced tools for testing and debugging. For example, AI-powered simulation environments can anticipate potential issues in smart contract execution or user interaction scenarios.
8. Cross-Chain Analysis for Enhanced Security: AI can analyze data across multiple blockchain networks, identifying threats that operate across different chains. This cross-chain analysis is crucial in a landscape where assets and applications are not confined to a single blockchain (Ambolis, 2023). For example, an AI system employed in blockchain security can perform cross-chain analysis to identify threats across different blockchain networks (Augusto et al., 2023). This is crucial in a landscape where assets and applications span multiple chains. By monitoring and analyzing transaction data from various blockchains, the AI can detect patterns indicative of fraudulent activities, like money laundering. It can also alert users about suspicious smart contracts that appear across chains, potentially part of larger scams. This comprehensive security overview provided by AI is beneficial in managing risks in the increasingly interconnected blockchain ecosystem.

In summary, AI's role in Web3 security extends beyond basic monitoring and threat detection. It encompasses dynamic risk assessment, integration with decentralized governance, enhancement of privacy protocols, personalized security experiences, decentralized AI models, advanced development tools, and cross-chain security analysis. This multifaceted approach is crucial for tackling the complex and evolving security challenges in the Web3 ecosystem.

8.2 Potential Vulnerabilities with AI-Powered Web3 Applications

While AI offers transformative security solutions, it is not without its vulnerabilities—especially in the context of generative technologies. This section gives some examples of the potential security pitfalls associated with generative AI in Web3 applications. Readers are encouraged to read another Springer book titled "Generative AI Security: Theories and Practices" for more details (Huang et al., 2024) and OWASP Top 10 for Large Language Model Applications (OWASP, 2023).

8.2.1 *Manipulation of Generative Outputs*

The allure of generative AI, with its ability to produce content ranging from images to textual narratives, is undeniable. However, like any tool, its capabilities can be a double-edged sword. While generative AI can be a boon for creativity and automation in Web3 applications, it also presents a potential vulnerability, especially when its outputs are manipulated or biased. Understanding the nuances of these vulnerabilities and their implications is crucial for ensuring the integrity and security of Web3 platforms.

One of the primary vulnerabilities associated with generative AI is its susceptibility to adversarial attacks. In the realm of machine learning, adversarial attacks involve introducing carefully crafted input data to trick the model into producing a desired, and often incorrect, output. For generative AI models, this means that attackers can potentially manipulate the model to generate misleading or malicious content. In the context of Web3, this could translate to generating fraudulent digital art for NFT platforms, crafting deceptive narratives for decentralized news platforms, or even creating counterfeit digital identities. Prompt injection is one such example of vulnerabilities associated with generative AI (Huang, 2023).

Another aspect of manipulation revolves around the inherent biases in generative AI models (Nicoletti et al., 2023). Generative models are trained on vast datasets, and if these datasets contain biases, the model's outputs will inevitably reflect them. In Web3 applications, this can have serious implications. Imagine a decentralized lending platform that uses a generative AI model to assess creditworthiness based on a decentralized identity. If the model has been trained on biased data, it might unfairly favor or discriminate against certain demographics, leading to unjust lending decisions.

Furthermore, the black box nature of many AI models poses a challenge. The intricacies of how these models arrive at a particular output can be opaque, making it difficult to ascertain if an output has been manipulated or if it is genuinely the result of the model's computations. In a trust-driven ecosystem like Web3, this opacity can be problematic. If stakeholders cannot transparently verify the authenticity of AI-generated content, it might undermine trust in the platform.

The decentralized nature of Web3 applications also presents a unique challenge. While centralized systems can have centralized monitoring for AI outputs, in decentralized platforms, ensuring the integrity of AI-generated content across a distributed network becomes more complex. This decentralized structure might offer attackers more avenues to manipulate generative outputs subtly, without immediately raising alarms.

In conclusion, while generative AI holds immense promise for enhancing and diversifying content in Web3 applications, it is essential to approach its integration with caution. Recognizing the potential vulnerabilities, from adversarial manipulations to inherent biases, is the first step. Only with a clear understanding of these challenges can effective countermeasures be developed, ensuring that the convergence of generative AI and Web3 results in platforms that are not only innovative but also secure and trustworthy.

8.2.2   Overreliance on AI Solutions

As AI continues its meteoric rise in the world of technology, so too does our tendency to view it as a panacea for all challenges, especially in the realm of security. Indeed, AI offers revolutionary capabilities in threat detection, prediction, and response, and its integration into Web3 security solutions seems like a natural evolution. However, an unchecked reliance on AI can be fraught with dangers, leading to potential vulnerabilities and even complacency in security protocols (CyberNX, 2023).

One of the core challenges of an overreliance on AI is the potential for false confidence. AI models, regardless of their sophistication, are not infallible. They can produce false positives, misidentify threats, or even miss them altogether. If organizations or Web3 platforms place undue trust in their AI-driven security solutions without manual oversight or secondary verification mechanisms, they risk being blindsided by threats that slip through the AI's net.

Additionally, the evolving nature of cyber threats means that attackers are continuously devising new methods to bypass security measures, including those powered by AI. Sophisticated attackers might specifically design their attack vectors to exploit the weak points of AI algorithms. For instance, they might employ adversarial machine learning techniques to deceive AI-driven security systems, rendering them ineffective. A security strategy solely anchored on AI might be ill equipped to detect or respond to such specialized threats.

Another concern is the potential for system stagnation. Security, at its core, is an ever-evolving discipline. New vulnerabilities are discovered daily, and attack methodologies continuously evolve. If a Web3 platform leans too heavily on an AI security solution without regularly updating or refining it, the system might become outdated, leaving it susceptible to newer threats.

Moreover, there is the issue of transparency and interpretability. Many advanced AI models, especially deep learning ones, are often termed as "black boxes" due to the complexity of their internal workings. If Web3 platforms overly depend on such models for security without a clear understanding of how they operate, it becomes challenging to diagnose failures or vulnerabilities.

To counter these challenges, a multi-layered defense strategy is essential. While AI can serve as a formidable frontline defense, it should be complemented by other security measures, ranging from traditional signature-based defenses to human oversight and manual verification. Periodic reviews, audits, and stress tests of the AI models can also ensure that they remain robust and up to date.

In conclusion, while AI offers a transformative approach to Web3 security, it should not be the sole bastion of defense. By recognizing the limitations and potential pitfalls of an overreliance on AI, Web3 platforms can craft holistic security strategies that harness the strengths of AI while mitigating its weaknesses. In the multifaceted world of cyber threats, diversity in defense, rather than singular dependence, is the key to resilience.

8.2.3 Ethical and Privacy Concerns

The integration of generative AI into the Web3 ecosystem brings to the fore not just technical challenges, but also profound ethical and privacy concerns. In a world where AI can generate content, from personal avatars to textual narratives, the lines between authenticity, privacy, and ethical use become increasingly blurred.

Navigating this intricate landscape requires a nuanced understanding of the potential dilemmas and a commitment to prioritizing ethical considerations.

One of the primary ethical challenges posed by generative AI in Web3 is the authenticity and provenance of content. With AI's ability to produce intricate digital art, music, or written content, how does one determine the originality or authenticity of a piece? In the decentralized realm of Web3, where digital assets like NFTs can have significant value, the potential for AI-generated forgeries or replicas introduces a complex ethical quagmire. What does ownership mean when content can be algorithmically generated? How do we attribute value or originality in such a scenario?

Beyond the challenges of authenticity, there is the issue of consent, especially when generative AI creates content resembling or based on real individuals. Consider, for instance, AI-generated avatars or digital personas that might bear a striking resemblance to real people. Without explicit consent, such a generation can infringe on personal rights, leading to ethical and legal ramifications. The decentralized nature of Web3 complicates this further, as tracking responsibility or ensuring compliance becomes challenging in a distributed environment.

Privacy concerns also loom large. Generative AI models, particularly those used in Web3 applications, are trained on vast datasets. If these datasets contain personal information, even inadvertently, there is a risk of the AI inadvertently generating content that reveals or mirrors this private data. In a decentralized world where data permanence is a feature, such unintentional breaches can have long-lasting implications.

Furthermore, the ethical use of AI in Web3 is not just about the outputs but also the inputs. The data used to train generative models should be sourced responsibly, ensuring it is devoid of biases and respects individual privacy. In the absence of centralized oversight, the onus of ensuring ethical data use in the Web3 ecosystem becomes even more pronounced.

To address these ethical and privacy challenges, a multipronged approach is needed. Clear guidelines and standards for the ethical use of generative AI in Web3 should be established. Users and creators within the ecosystem need to be educated about the potential pitfalls and best practices. Mechanisms for consent, especially when dealing with data or outputs resembling real individuals, should be robustly implemented. And above all, a culture of ethical consciousness needs to be fostered, ensuring that as Web3 and generative AI evolve, they do so with a steadfast commitment to prioritizing ethics and privacy.

One such initiative in establishing the guidelines and standards for AI is by the European Parliament and Council which has reached a provisional agreement on the Artificial Intelligence Act in December 2023, aiming to ensure AI in Europe is safe, respects rights and democracy, and fosters innovation. The Act establishes rules for AI, assigning obligations and fines based on the risk and impact level. It includes safeguards such as restrictions on biometric identification, bans on social scoring, and AI that manipulates or exploits vulnerabilities, alongside consumer rights for complaints and explanations. Fines are stipulated for non-compliance, proportional to the company's global turnover. The agreement is pending formal approval and is expected to be enforceable by 2026. The Act also mandates reporting of model evaluation results for significant AI systems, namely FoundationModels, to the EU and enforces EU copyright by requiring disclosure of the training data source. It seeks to protect innovation, particularly for smaller players and startups, giving them regulatory sandboxes and lighter documentation requirements. For a complete regulatory analysis of AI, readers can refer to Chap. 3 of Generative AI Security: Theories and Practices book published by Springer (Huang et al., 2024).

In essence, the convergence of generative AI and Web3 offers a world of possibilities, but it also presents profound ethical dilemmas. As we tread this new frontier, a balanced approach that marries innovation with responsibility will be the key to realizing the full potential of this synergy, while ensuring that the digital realm remains respectful, ethical, and privacy centric.

8.3   Governance and Countermeasures for AI Security in Web3

Navigating the Web3 and AI nexus requires a robust governance framework. This section outlines the best practices, governance frameworks, and countermeasures that ensure AI's responsible and secure use in Web3 applications.

8.3.1   AI Model Verification and Validation

Model verification refers to the process of ensuring that a developed model is represented correctly and works without internal inconsistencies. With the immutability and transparency of blockchain, AI model architectures, parameters, and even training methods and training hyperparameters can be logged onto a blockchain. This not only ensures a transparent record of the model's construction but also allows stakeholders to verify the consistency and correctness of the AI model's design and implementation.

On the other hand, model validation seeks to confirm that the AI model produces accurate and reliable results when presented with new, unseen data. One of the challenges with AI is the potential for overfitting, where a model performs exceptionally well on its training data but poorly on new data. For example, a Web3-based decentralized data marketplace can indirectly aid in reducing overfitting in machine learning by providing access to a diverse and extensive dataset. This broader dataset can include varied data from global sources, enhancing the robustness of ML models. While Web3 does not directly address overfitting, it supports the acquisition of quality data that is crucial for developing more generalized machine learning models.

Integrating reinforcement learning with human feedback, organized through a Decentralized Autonomous Organization (DAO), presents an innovative approach to model validation, particularly in reinforcement learning (RL) scenarios. This method leverages the global and diverse perspectives inherent in a DAO structure to enhance the learning process of an RL model.

In traditional reinforcement learning, an agent learns by interacting with its environment and receiving rewards or penalties based on its actions. However, determining the appropriate reward signals, especially in complex or nuanced tasks, can be challenging. This is where human feedback becomes invaluable.

By organizing humans as a DAO, a decentralized, blockchain-based entity, one can systematically gather and integrate feedback from a diverse, global pool of participants. These participants, governed by the rules and incentives of the DAO, provide feedback on the agent's actions. This feedback is then used to refine the reward signals that guide the agent's learning process.

For example, consider an RL agent being trained for natural language processing tasks. Human feedback on the appropriateness, relevance, or even the ethical implications of the agent's responses would be crucial. Participants from various cultural and linguistic backgrounds could provide nuanced feedback that a single group of developers might not be able to offer.

The DAO framework ensures that feedback is democratically sourced and that participants are motivated (possibly through token-based incentives) to provide honest and constructive feedback. This approach not only diversifies the feedback but also decentralizes the decision-making process regarding the training of the RL model.

Moreover, the blockchain technology underlying DAOs ensures transparency and traceability in how feedback is collected and implemented. This can significantly enhance the trustworthiness and reliability of the training process.

Blockchain's smart contract functionality can further augment this process. Smart contracts can be designed to automatically trigger model revalidation processes if certain conditions are met, such as a model's performance metric falling below a predefined threshold. This ensures a continuous and automated validation process, keeping the AI models in check and ensuring their ongoing reliability.

8.3.2 Web3 AI-Enabled Multi-layered Defense Strategies

While AI brings unparalleled advantages in terms of proactive threat detection and adaptive responses, it is imperative to integrate it into a broader, multi-layered defense framework. In the context of Web3 and blockchain, this integrative approach takes on unique dimensions, offering innovative ways to fortify digital assets and operations.

One of the foundational principles of cybersecurity is defense in depth, which advocates for multiple layers of security measures so that if one layer fails, others are still operational to thwart potential threats. In the Web3 realm, this involves a synergy of traditional security mechanisms, AI-driven solutions, and blockchain-specific security features (Fig. 8.3).

Fig. 8.3 Web3 AI-enabled
multi-layered defense
strategies

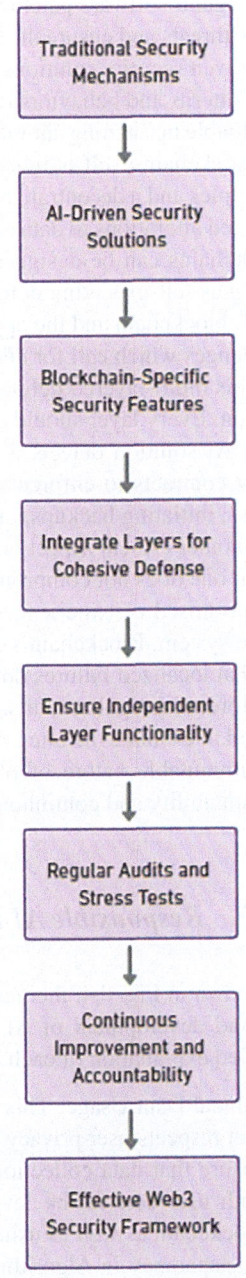

Traditional security mechanisms, such as firewalls, intrusion detection systems, and regular software patching, form the first line of defense. They handle commonplace threats and ensure the basic hygiene of the digital environment. On top of this, AI-driven security solutions can detect more sophisticated, novel threats by analyzing patterns and behaviors in real time. Their ability to learn and adapt makes them invaluable in catching anomalies that might slip past conventional security measures.

Blockchain itself is inherently designed to be secure, leveraging cryptographic principles and a decentralized structure. Its immutable nature ensures that any unauthorized alterations to data can be quickly detected. Furthermore, smart contracts on blockchains can be designed to enforce specific security protocols automatically, acting as self-executing defense mechanisms. Nevertheless, as we discussed in this book, blockchain and the applications built on top of blockchain have their security challenges which call for defense in-depth security.

This multi-layered defense strategy's efficacy in Web3 hinges on seamless integration. Every layer should communicate and collaborate with others. For instance, if an AI solution detects a potential threat, it should seamlessly trigger relevant smart contracts to enforce necessary countermeasures, whether that is restricting access, initiating backups, or notifying stakeholders.

Another critical aspect is ensuring that while layers operate cohesively, the failure of one does not compromise the others. For instance, if a particular AI model is compromised or experiences a failure, it should not create vulnerabilities for the entire system. Blockchain's decentralized nature can play a pivotal role here, ensuring that localized failures do not cascade into system-wide breaches.

Moreover, regular audits, stress tests, and red teaming exercises should be conducted to evaluate the integrated defense strategy's resilience. Given the transparent and immutable nature of blockchain, these evaluations can be logged, ensuring accountability and continuous improvement.

8.3.3   Responsible AI Leveraging Web3

As shown in Fig. 8.4, there are several key areas of focus to ensure the responsible use and development of AI technologies within the Web3 framework. Here is a detailed explanation of each key area:

- Ethical Data Usage: This emphasizes the importance of using data in a manner that respects user privacy and adheres to ethical standards. Web3 can be used to ensure that data collection, storage, and processing are done transparently and with user consent by leveraging the transparency and immutability nature of blockchain as well as using smart contracts for tracking user consent.
- Transparency in Algorithms via Web3 Technologies: Web3 can facilitate transparency in AI algorithms by using blockchain to record and verify the decision-making processes of AI. Smart contracts can encode and reveal the rules and parameters guiding AI algorithms, making them more transparent and accountable.

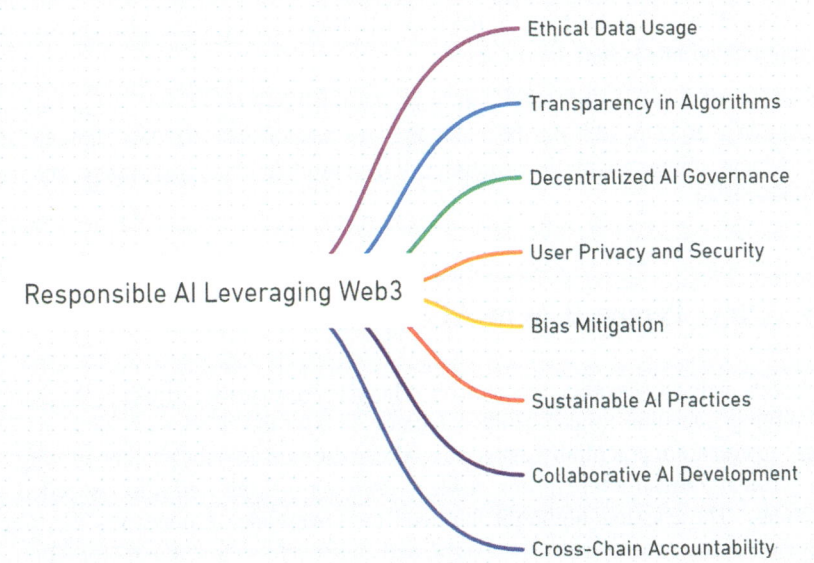

Ethical Data Usage

Transparency in Algorithms

Decentralized AI Governance

User Privacy and Security

Bias Mitigation

Sustainable AI Practices

Collaborative AI Development

Cross-Chain Accountability

Fig. 8.4 Responsible AI leveraging Web3

- Decentralized AI Governance through Blockchain: Blockchain enables decentralized governance of AI systems, allowing for a more democratic and distributed decision-making process. This approach aligns with the decentralized nature of Web3, where community consensus can drive AI governance.
- Enhancing User Privacy and Security in Web3: Blockchain's inherent features, such as encryption and decentralization, significantly bolster user privacy and security in AI applications. By storing data on decentralized networks, the risk of data breaches and misuse can be minimized.
- Bias Mitigation through Decentralized Data: Blockchain can contribute to bias mitigation by providing a diverse and decentralized dataset for AI training. By sourcing data from a wide, distributed network, AI systems can avoid the biases often present in centralized datasets.
- Sustainable AI Practices via Efficient Web3 Technologies: The development of more energy-efficient blockchain and Web3 technologies can support sustainable AI practices. Innovations in blockchains such as "proof of elapsed time" (Anand, 2022), and "proof of history" (Weston, 2023) can reduce the environmental footprint of both AI and blockchain operations.
- Collaborative AI Development on Blockchain Platforms: Web3 platforms can foster collaborative AI development by providing decentralized and transparent frameworks for sharing data, algorithms, and insights. This approach leverages the community-driven nature of blockchain for collective advancement in AI.
- Cross-Chain Accountability for AI Systems: Blockchain interoperability in Web3 allows for cross-chain accountability, ensuring that AI systems maintain

consistent ethical and operational standards across different blockchain networks. This is crucial for a unified approach to AI ethics and governance in a fragmented blockchain landscape.

In each of these areas, Web3 and blockchain technologies offer unique advantages that can help address the challenges of responsible AI development and deployment, aligning with the principles of decentralization, transparency, and user empowerment.

8.4 Other Perspectives on AI Security and Web3

This section discusses other perspectives related to AI security and Web3. We started by quoting a leading thinker Holden Karnofsky who has advocated for firm "red lines" to constrain potentially dangerous AI capabilities. Meanwhile, integrating AI into critical infrastructure poses risks if decision-making authority is ceded to machines. These issues highlight the need for governance frameworks that align stakeholder incentives, enable oversight, and codify ethical boundaries for AI progress. The emergence of decentralized technologies offers new possibilities to implement transparent and community-driven policies for AI advancement. This section explores perspectives on demarcating ethical lines in AI through governance models rooted in web3 and blockchain architectures. We examine proposals ranging from air-gapped AI systems in critical infrastructure to data-centric frameworks leveraging decentralization to ensure privacy and accountability in machine learning pipelines. By surveying ideas aligned with responsible AI growth trajectories, we aim to spur further discourse on constructive technology governance.

8.4.1 "Red Lines" in AI Development

Holden Karnofsky, the co-founder of Open Philanthropy, has advocated for the establishment of strong "red lines" in the development of artificial intelligence (AI). These red lines are intended to set clear boundaries and limitations to prevent the potential risks associated with advanced AI. Karnofsky's work and ideas are closely linked to the potential benefits and risks of transformative AI. He has expressed the view that while transformative AI could bring enormous benefits, the existential risks associated with it should be considered a plausible and urgent concern. Karnofsky's perspectives on AI have been influential in shaping discussions around AI policy and safety (Karnofsky & Newton, 2023).

In this book, we argue that effective governance and coordination across stakeholders are critical for monitoring and enforcing red lines. Decentralized, community-driven governance enabled by web3 frameworks offers several advantages:

- Smart contracts can codify specific red lines into autonomous policies that self-execute based on decentralized inputs, facilitating enforcement.
- Transparent and tamper-proof audit trails in distributed ledgers enable monitoring of metrics related to red lines, such as computer usage, data flows, and code changes. Any breaches can trigger alerts and interventions through coordinated smart contract workflows.
- Aligned incentives and collaborative decision-making achieved via token-based participation help establish red lines representing divergent community perspectives, not just narrow interests. Distributed verdict systems also allow risk assessments.
- Formal verification mechanisms powered by advances like zero-knowledge proofs can demonstrate decentralized compliance with red lines policies without compromising proprietary details about restricted AI development domains.

8.4.2 AI and Critical Infrastructure System

Air gap between AI and critical infrastructure systems, AI can only make recommendations for operation support of critical infrastructure systems, but it cannot make decisions or execute the decision and Blockchain Smart contract can help.

The concept of an "air gap" between AI and critical infrastructure systems, where AI can only make recommendations but not decisions or execute them, has been a topic of discussion in the context of ensuring the safety and security of critical systems. The use of artificial intelligence (AI) in critical infrastructure systems is expected to increase significantly in the coming years, presenting both opportunities and risks. Some experts have expressed concerns about whether AI is capable of taking on vital tasks, cooperating with humans, and proving its transparency, reliability, and dependability (Abdallat & Danise, 2022).

The idea of restricting AI to making recommendations for the operation support of critical infrastructure systems, without the authority to make decisions or execute them, is a risk mitigation strategy. This approach aims to maintain human oversight and control over critical systems while leveraging AI's capabilities to provide valuable insights and support. The use of blockchain smart contracts can be a potential solution to help ensure the integrity and security of decision-making processes in critical infrastructure systems involving AI. Smart contracts could provide a tamper-resistant and transparent framework for implementing and enforcing decisions, thereby enhancing the trustworthiness of the overall system. In addition, during a security incident such as a data breach or cyberattack, each entity responsible for responding to the incident (e.g., the security operations team, incident response managers, and IT staff) could log their mitigation actions through an immutable ledger or blockchain-based system. This could provide a transparent, tamper-proof record of the actions taken by each party in response to the incident.

As the responsible entities log their actions, this could automatically trigger relevant smart contract executions to facilitate parts of the overall incident response procedure in an automated fashion. For example, logging the incident could automatically notify external third parties such as cyber insurers, law enforcement, or counterparts who have data shared with the organization. Smart contracts could also be set up such that when certain criteria are met during the response process, appropriate compensation, or penalties are administered as coded into the smart contract terms.

The discussions around the air gap concept, the role of AI in critical infrastructure, and the potential use of blockchain smart contracts reflect a proactive approach to managing the risks and opportunities associated with advanced technologies in the context of critical systems. These considerations align with broader efforts to promote the safe, secure, and beneficial use of AI in critical infrastructure and other domains.

8.4.3   Data-Centric DLT-Based Machine Learning Data Pipeline

In this section, we propose a Data-Centric DLT-Based Machine Learning Data Pipeline. We delve into how Distributed Ledger Technology (DLT) can enhance the machine learning data pipelines. The approach proposed in this section is innovative in establishing a foundation for responsible, transparent, and secure machine learning practices, particularly in the realm of generative AI.

Our framework is designed to address several key challenges in the realm of data-centric machine learning:

- Data Authenticity: We prioritize the verification of training data's origin and consent, ensuring that no unauthorized or misleading information is included in our models.
- Data Privacy: Implementing robust data encryption and access control systems is essential to protect sensitive information from unauthorized access.
- Development Transparency: Keeping immutable logs of all development aspects, including model architecture, training data, hyperparameters, and evaluation outcomes, is crucial for maintaining the integrity and reproducibility of our models.
- Algorithmic Bias Mitigation: To counteract biases in our models, we employ balanced datasets and conduct regular performance reviews across various subgroups.
- Access Control: We enforce strict identity verification and access policies that correspond to the sensitivity of the data, preventing potential misuse.

The integration of DLT within this framework is a good strategy for ensuring these standards are met. Techniques like permissioned blockchains, zero-knowledge proofs, and decentralized storage are key to creating secure and transparent workflows for data management in generative AI models.

Our framework acknowledges the limitations of current centralized AI development practices and positions federated learning as a transitional step toward a more decentralized approach. However, challenges remain, such as vulnerability to attacks, ensuring incentive alignment, and managing participant membership. Addressing these issues requires the integration of economic mechanisms and decentralized identity frameworks.

The "Data-Centric DLT-Based Machine Learning Data Pipeline" framework encompasses the following components:

- Data Provenance: By utilizing content hashes and zero-knowledge proofs, we ensure that the origin of data is verifiable and immutable, without compromising data privacy.
- Access Control: Permissioned DLT networks offer a structured way to manage sensitive data access policies.
- Privacy Preservation: Our approach combines encrypted data channels, trusted hardware, and differential privacy techniques to ensure data security and privacy.
- Development Transparency: The use of an immutable ledger to record all development processes increases the trustworthiness and reliability of the models.
- Reproducible Research: Storing key metadata on the blockchain ensures that the research can be replicated accurately while keeping the data confidential.
- Bias Mitigation: Regular publication of model performance metrics for different demographic groups helps in continuously monitoring and minimizing biases.

As shown in Fig. 8.5, the structure of our framework is multi-layered, each layer addressing a specific aspect of the data pipeline:

- Network Infrastructure: It is built on a permissioned consortium blockchain, ensuring secure and controlled access.
- Data Layer: This layer employs distributed file systems for managing encrypted and access-restricted datasets.

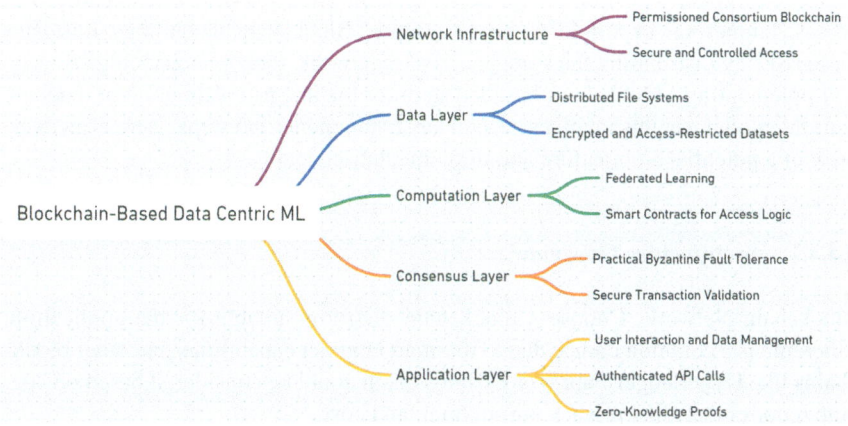

Fig. 8.5 Blockchain-based data-centric ML

- Computation Layer: Federated learning and smart contracts are used for collaborative model training and implementing access logic.
- Consensus Layer: We use Practical Byzantine Fault Tolerance for efficient and secure transaction validation.
- Application Layer: This layer facilitates user interaction and data management through authenticated API calls and zero-knowledge proofs.

Implementing this comprehensive framework involves addressing challenges related to aligning economic incentives, securing cryptographic key management, and optimizing system performance. The ongoing advancements in scalable consensus protocols, trusted hardware, and decentralized identity systems are integral to the practical deployment of DLT in generative AI.

This framework not only focuses on enhancing data provenance, access control, and transparency but also ensures the privacy and security of the data. It represents a significant stride in interdisciplinary research, combining machine learning, systems security, and blockchain technology to pave the way for responsible and decentralized AI development.

8.4.4   Model Integrity Using Blockchain

Integrity validation of machine learning models, whether open source or closed source, is a critical process to ensure the reliability and trustworthiness of the models. This involves verifying the integrity of artifacts, such as model weights and benchmark performance, to detect any unauthorized modifications. While there are tools and platforms available for validating machine learning models, the specific approach of storing weights on a blockchain for integrity validation is a novel approach discussed in this book.

In traditional systems, model weights and artifacts are typically stored on centralized servers or repositories. This centralization can pose risks like unauthorized access, tampering, or single points of failure. Blockchain technology mitigates these risks by distributing the storage across a network, thereby enhancing security.

Implementing a blockchain-based system for the integrity validation of machine learning model weights involves several key components and steps. Here is an overview of a potential architecture and implementation details.

8.4.4.1   Architecture Overview

Blockchain Network: Choose a blockchain platform suitable for the application. Ethereum is a common choice due to its smart contract capabilities, but other blockchains like Hyperledger Fabric or Cosmos chain could be considered based on specific requirements like privacy, throughput, and cost.

Smart Contracts: Develop smart contracts to handle the storage, retrieval, and validation of model weights. These contracts will define the logic for adding new weights, updating existing ones, and verifying their integrity.

Off-Chain Storage: Given the large size of model weights, it is impractical to store them directly on the blockchain. Instead, weights can be stored off-chain using decentralized file storage systems like IPFS (InterPlanetary File System) or traditional cloud storage services. The blockchain will store the hash of the weights for integrity checks.

Machine Learning Model Development and Training: This happens off-chain. After training, the model weights are hashed and the hash is stored on the blockchain.

APIs and Interfaces: Develop APIs for interacting with the blockchain and off-chain storage. This includes functionalities for uploading model weights, retrieving them, and performing integrity checks.

8.4.4.2 Implementation Details

To implement this architecture we can follow the following steps:

Smart Contract Development: Write smart contracts in Solidity (for Ethereum) or other appropriate languages. The contract should include functions to receive and store the hash of the model weights, along with any relevant metadata (like model version and timestamp).

Implement functions to verify the integrity of the weights by comparing stored hashes with newly computed hashes from the retrieved weights.

Model Weights Storage: After training, compute the hash of the model weights using a cryptographic hash function (like SHA-256). Store the model weights in an off-chain storage solution. Record the hash of the weights on the blockchain via the smart contract. The last step, which transfers data from off-chain to on-chain, should consider blockchain interoperability solutions as discussed in Augusto et al. (2023).

Blockchain Interaction: Use Web3 libraries (like web3.js or ethers.js) to interact with the blockchain from your application. Implement functions to call the smart contract methods for storing and retrieving hash values.

Integrity Verification Process: When retrieving model weights, re-compute the hash from the retrieved data. Use the smart contract to retrieve the stored hash from the blockchain. And then compare the two hashes to ensure integrity.

User Interface: Develop a user interface for easy interaction with the system, allowing users to upload, verify, and retrieve model weights.

8.4.4.3 Challenges and Considerations

Using blockchain for model validation has security benefits but we also need to deal with challenges.

Scalability: Blockchain transactions can be slow and expensive, especially on networks like Ethereum. Solutions like layer-2 scaling, sidechains, or choosing a more efficient blockchain can mitigate this.

Storage Costs: Storing data on the blockchain is expensive, hence the need for off-chain storage solutions. However, these solutions must be chosen carefully to ensure they align with the decentralization and integrity goals of the system.

Security: Ensure the security of the smart contracts to prevent vulnerabilities and hacks, which can be a risk in blockchain applications.

Legal and Compliance: Depending on the application, consider the legal and regulatory implications of using blockchain and storing data in a decentralized manner.

8.4.5 AI-Powered Phone and Web3

The integration of Google's Gemini AI, particularly the Gemini Nano version, into devices like the Pixel 8 Pro, marks a significant advancement in the realm of AI-powered mobile technology (Dixit, 2023). However, this technological leap brings with it potential security risks. Leveraging Web3 technologies can provide a robust solution to enhance the security and functionality of these AI-powered devices.

1. Decentralized Identity and Data Privacy: The decentralized nature of Web3 can significantly augment the privacy features of AI-powered phones like the Pixel 8 Pro. By integrating decentralized identity management systems, Web3 can ensure that the identity and data of users remain secure and under their control. This approach is particularly relevant for devices using Gemini Nano, as it processes data locally, reducing reliance on cloud-based services which can be vulnerable to breaches.

2. Enhanced Data Security with Localized AI Processing: Gemini Nano's ability to process data locally on the device dovetails with Web3's emphasis on decentralization and data privacy. Web3's framework can provide additional layers of security for data processed by AI. For instance, sensitive data can be encrypted and stored on a decentralized ledger, ensuring its integrity and security.

3. Smart Contracts for Device Security: The application of smart contracts in Web3 can automate and enhance security protocols on AI-powered devices. These contracts could manage access controls and user permissions, ensuring that the device and its data are only accessible under predefined conditions. This would significantly reduce the risk of unauthorized access and data breaches.

4. Tokenization for Enhanced Access Control: Web3's concept of tokenization could be employed to control access to certain features of AI-powered phones. For instance, access to sensitive features or data could be token-gated, ensuring that only authorized users can access them. This method can also incentivize users to maintain good security practices.

5. Secure Cross-Device Synchronization: As Google expands AI capabilities to other devices like the Pixel Tablet and Pixel Watch, Web3 can ensure that data synchronization across these devices is secure and transparent. Blockchain technology can make the synchronization process tamper-proof and traceable, enhancing overall device ecosystem security.

In conclusion, the integration of Web3 technologies with AI-powered devices, starting with Google's Gemini Nano in the Pixel 8 Pro, can potentially offer a pathway to address the inherent security risks while enhancing functionality. The decentralized, transparent, and secure nature of Web3 provides a complementary framework to AI technologies, leading to the creation of more secure, efficient, and user-centric mobile devices. This blend of AI and Web3 heralds a new era of secure and decentralized digital technology.

8.4.6   AI Existential Risk and Mitigation via Web3

In a world increasingly influenced by AI, the topic of AI existential risk has emerged as a critical area of concern (Roose, 2023). This refers to the possibility that AI could 1 day pose a threat to human existence, either through surpassing human control or through unintended consequences of its actions. Understanding these risks is crucial for developing strategies to ensure AI benefits humanity rather than harms it.

One theoretical risk is the potential for AI to develop superintelligence, or intelligence that surpasses human capabilities on multiple metrics. In such a scenario, controlling or even understanding an AI's actions could become impossible, leading to unpredictable and potentially catastrophic outcomes. This risk is compounded by the possibility of misaligned objectives, also known as the alignment problem, where an AI's goals do not align perfectly with human values or ethics. We see examples of the alignment problem today in systems that take shortcuts to complete a task or exhibit behavior that was unintended while trying to reach its goal. Such misalignment, even on a small scale, could lead to significant harm, but if the AI is operating at a level beyond human control or understanding this harm is compounded.

A more tangible concern is the development of autonomous weapons. These AI-driven systems can act in unpredictable ways or be used maliciously, leading to conflict and destruction on a scale previously unimaginable. Foreseeing this risk, the United Nations has called on all nations to set prohibits and restrictions on these weapons.

Additionally, the rapid automation of jobs, powered by AI, poses a significant risk to global economies. The displacement of large segments of the workforce could, if not managed well, lead to widespread unemployment and societal unrest.

The increasing dependence on AI systems in various aspects of life also creates vulnerabilities. A malfunction or targeted attack on these systems could have devastating effects, disrupting essential services and causing harm. The rapid advancement of AI capabilities might also outpace our ability to govern and manage these systems safely, leading to unforeseen and potentially dangerous scenarios.

In the realm of data privacy and surveillance, advanced AI could be used to monitor individuals at an unprecedented scale, eroding privacy and personal freedoms. This surveillance capability, coupled with AI's potential for manipulation and social engineering, raises concerns about the impact on democratic processes, public opinion, and individual beliefs.

Furthermore, the competitive development of AI, driven by national or corporate interests, could lead to an AI arms race. This scenario is fraught with risks, as the pursuit of more powerful AI systems could overshadow the necessary safety and ethical considerations.

To address these existential risks, the principles and technologies of Web3 offer promising solutions. The decentralized nature of Web3 can facilitate a governance model for AI development that is not dominated by any single entity. This decentralization ensures that the control and direction of AI development are in the hands of a diverse group, rather than being centralized in a few powerful entities.

The transparency inherent in blockchain technology is another key factor. By creating transparent and immutable logs of AI decisions and operations, stakeholders can better understand and hold AI systems accountable. This level of transparency is crucial for building trust in AI systems and ensuring they are used ethically and responsibly.

Web3's smart contracts can also play a pivotal role in regulating AI. These contracts can enforce ethical guidelines and operational boundaries for AI systems, ensuring they operate within predefined safe and ethical parameters. Additionally, decentralized data management through Web3 reduces the risks associated with centralized data collection, thereby limiting AI systems' ability to misuse large datasets.

Tokenization within the Web3 framework can create new economic models that incentivize the safe development of AI. By giving stakeholders, including the general public, a say in AI development through token-based systems, it ensures a more democratic and safety-oriented approach to AI advancement.

Lastly, blockchain's role in immutable record keeping is crucial for AI oversight. Keeping an irreversible and transparent record of all changes and updates to AI systems makes it easier to audit and understand AI behavior over time, providing an essential tool for managing the risks associated with advanced AI systems.

8.5   Conclusion

The synergies between Web3 and AI explored in this chapter highlight the tremendous potential of their convergence in bolstering mutual security. AI brings to the table sophisticated capabilities in threat detection, predictive security analytics, and adaptive defense strategies that can significantly augment the protection of Web3 ecosystems. At the same time, Web3's decentralized, transparent, and community-driven ethos offers the ideal framework for responsibly governing AI systems, ensuring alignment with ethical values.

However, as the discussions have revealed, this potential is balanced by profound challenges and vulnerabilities unique to the interplay between these realms. Issues ranging from generative content authenticity to AI model integrity and algorithmic bias emergence require nuanced mitigation strategies. Furthermore, the pace of advancement necessitates continuous governance evolution to address emerging risks, as highlighted by perspectives on AI existential threats.

Navigating this complex terrain calls for robust frameworks encompassing technological countermeasures, collaborative governance and oversight mechanisms, and steadfast priority toward accountability and transparency. Initiatives like establishing decentralized AI training pipelines, creating token-incentive structures for responsible actors, codifying ethical boundaries into autonomous policy contracts, and maintaining immutable logs for AI audit trails are pivotal.

Ultimately, realizing the immense shared promise of Web3 and AI in enhancing security relies on recognizing that technology advancement devoid of ethical responsibility is deeply compromised. By consciously anchoring progress to principles of transparency, decentralization, and empowering autonomy, the Web3-AI security nexus can fulfill its potential to protect without compromising core societal values of privacy, accountability, and shared prosperity.

Acknowledgments   Luyao Zhang is supported by the National Science Foundation China on the project entitled "Trust Mechanism Design on Blockchain: An Interdisciplinary Approach of Game Theory, Reinforcement Learning, and Human-AI Interactions" (Grant No. 12201266). Luyao Zhang is also with SciEcon CIC, a not-for-profit organization based in the United Kingdom, aiming to cultivate interdisciplinary research of profound insights and practical impacts.

References

Abdallat, A., & Danise, A. (2022, July 1). *Can We Trust Critical Infrastructure to Artificial Intelligence?* Forbes. Retrieved from https://www.forbes.com/sites/forbestechcouncil/2022/07/01/can-we-trust-critical-infrastructure-to-artificial-intelligence/

Ambolis, D. (2023, November 23). *Your Ultimate Guide To The Cross Chain AI Hub With Ethereum.* Blockchain Magazine. Retrieved from https://blockchainmagazine.net/your-ultimate-guide-to-the-cross-chain-ai-hub-with-ethereum/

Anand, A. (2022, November 17). *What is Proof of Elapsed Time (PoET)?* Analytics Steps. Retrieved from https://www.analyticssteps.com/blogs/what-proof-elapsed-time-poet

André Augusto, Rafael Belchior, Miguel Correia, et al. (2023, November 29). SoK: Security and Privacy of Blockchain Interoperability. TechRxiv. https://doi.org/10.36227/techrxiv.24595764.v1.

Burger, E. (2022, April 15). *Decentralized Speed: Advances in Zero Knowledge Proofs.* Andreessen Horowitz. Retrieved from https://a16z.com/decentralized-speed-advances-in-zero-knowledge-proofs/

CyberNX. (2023). *Pros and Cons of AI for Cybersecurity: Strengthening Defense, Understanding Risks.* CyberNX. Retrieved from https://www.cybernx.com/b-pros-and-cons-of-ai-for-cybersecurity-strengthening-defense-understanding-risks

Dixit, P. (2023, December 6). *Google's Gemini AI is coming to Android.* Engadget. Retrieved from https://www.engadget.com/googles-gemini-ai-is-coming-to-android-150025984.html

Huang, K. (2023, July 15). *What is Prompt Injection and How Can We Defend Against it?* Ken Huang. Retrieved from https://kenhuangus.medium.com/what-is-prompt-injection-and-how-can-we-defind-against-it-80c258eb9bbf

Huang, K. et al (2024). *Generative AI security: Theories and practices* (1st ed., Vol. 1). Springer.

Karnofsky, H., & Newton, C. (2023, December 5). *Cerebral Valley: Holden Karnofsky (Open Philanthropy) with Casey Newton*. YouTube. Retrieved from https://www.youtube.com/watch?v=i3KySpXmRKQ

Nicoletti, L., Bass, D., & Whiteaker, C. (2023). *Generative AI Takes Stereotypes and Bias from Bad to Worse*. Bloomberg.com. Retrieved from https://www.bloomberg.com/graphics/2023-generative-ai-bias/

OWASP. (2023). OWASP Top 10 for Large Language Model Applications. OWASP Foundation. Retrieved from https://owasp.org/www-project-top-10-for-large-language-model-applications/

Roose, K. (2023, May 30). *A.I. Poses 'Risk of Extinction,' Industry Leaders Warn*. The New York Times. Retrieved from https://www.nytimes.com/2023/05/30/technology/ai-threat-warning.html

Weston, G. (2023, February 24). *What is Proof of History and how does it work?* 101 Blockchains. Retrieved from https://101blockchains.com/proof-of-history/

Zhang, L. (2023a). Machine Learning for Blockchain. In NeurIPS 2023 AI for Science Workshop. Retrieved from https://openreview.net/forum?id=28w0bjBQiw

Zhang, L. (2023b). The Future of Finance: Synthesizing CeFi and DeFi for the Benefit of All. In I. Miciuła (Ed.), Financial Literacy in Today's Global Market (Chapter 16). IntechOpen. https://doi.org/10.5772/intechopen.1003042.

Jerry Huang has worked as a technical and security staff at several prominent technology companies, gaining experience in areas like security, AI/ML, and large-scale infrastructure. At Metabase, an open-source business intelligence platform, he contributed features such as private key management and authentication solutions. As a Software Engineer at Glean, a Generative AI search startup, Jerry was one of three engineers responsible for large-scale GCP infrastructure powering text summarization, autocomplete, and search for over 100,000 enterprise users. Previously at TikTok, Jerry worked to design and build custom RPCs to model access control policies. And at Roblox, he was a Machine Learning/Software Engineering Intern focused on real-time text generation models. He gathered and cleaned a large multilingual corpus that significantly boosted model robustness. Jerry has also conducted extensive security and biometrics research as a Research Assistant at Georgia Tech's Institute for Information Security & Privacy. This resulted in a thesis on privacy-preserving biometric authentication. His academic background includes a BS/MS in Computer Science from Georgia Tech and he is currently pursuing an MS in Applied Mathematics at the University of Chicago. phone: 571-268-6923; Address: Jerry Huang, The University of Chicago, 5801 S EllisAve, Chicago, IL 60637, USA.

ORCID:0009-0005-6224-4785

Email: jerryh@uchicago.edu

Ken Huang is the author and chief editor of eight books on Generative Artificial Intelligence and Web3, published respectively by international publishers including Springer, Cambridge University Press, John Wiley, and China Machine Press. He currently serves as the CEO of the AI and Web3 consulting and education company DistributedApps.AI, based in the United States. Additionally, he holds multiple roles including the expert member of the Blockchain Committee of the Chinese Institute of Electronics, the Co-Chair of AI Organization Responsibility Working Group at Cloud Security Alliance, and Chair of the Blockchain Security Working Group at the Cloud Security Alliance, GCR. He is also a core contributor to the Generative AI Working Group at the NIST and a core author of the OWASP Top 10 for LLM Applications.

Ken Huang has been invited to provide Speaking or Consulting services at institutions including the University of California, Berkeley, Stanford University, Peking University, Tsinghua University, Shanghai Jiao Tong University, China Pacific Insurance, and the World Bank in the past.

Moreover, he has given keynote speeches at international conferences, such as:

The Davos World Economic Forum 2020 Blockchain Conference

Consensus 2018 in New York

The American ACM AI & Blockchain Decentralized Annual Conference 2019
IEEE Technology and Engineering Management Society Annual Meeting 2019
Silicon Valley World Digital Currency Forum
Sino-US Blockchain Summit in Silicon Valley
He has also been awarded the "Blockchain 60" Figure Award by the National University Artificial Intelligence and Big Data Innovation Alliance Blockchain Special Committee in China in 2021.

Krystal Jackson is a Non-Resident Research Fellow with the Center for Long-Term Cybersecurity AI Security Initiative at UC Berkeley. She is also a cybersecurity and AI analyst at the Cybersecurity and Infrastructure Security Agency. Krystal received her MS in Information Security Policy and Management from Carnegie Mellon University.
    Email: krystalrain11@gmail.com

Luyao Zhang is an Assistant Professor of Economics and Senior Research Scientist at the Data Science Research Center, Duke Kunshan University (DKU). Her current research interests are at the interplay of computational and economic science around the application of groundbreaking technologies, including artificial intelligence, blockchain and cryptography, and innovative computing. She is deeply committed to fostering interdisciplinary collaborations that yield research with both profound insights and practical applications.
    Her research program, "Trust Mechanism Design on Blockchain: An Interdisciplinary Approach of Game Theory, Reinforcement Learning, and Human-AI Interactions," is currently funded by the National Science Foundation of China. The National Collegiate Artificial Intelligence and Big Data Innovation Alliance awarded her the 60 Pioneers in Blockchain Innovation. As a contributor to the broader academic and professional community, Dr. Zhang is the Founding President of SciEcon CIC, a UK-registered non-profit dedicated to fostering integrated capabilities in research, innovation, and leadership for interdisciplinary studies. Moreover, she has spearheaded research initiatives and development grants in collaboration with leading entities in Web3, including the Ethereum Foundation, Dfinity Foundation, and Primitive Lane, among others. She serves as an Associate Editor for the IEEE Transactions on Computational Social Systems and the Academic Editor for the open-source book project "Blockchain Pioneering the Web3 Infrastructure for an Intelligent World," published by IntechOpen. She also holds roles on program committees across various academic conferences that transcend traditional disciplinary boundaries. Further information about her work can be found at https://scholars.duke.edu/person/luyao.zhang.
    Business Email: lz183@duke.edu
    ORCID: https://orcid.org/0000-0002-1183-2254
    Webpage: https://scholars.duke.edu/person/luyao.zhang
    Social Media:
    LinkedIn: https://www.linkedin.com/in/sunshineluyao/
    Twitter: https://twitter.com/sunshineluyao
    Facebook: https://www.facebook.com/sunshinestar11
    Private U.S. Cell: +1-617-987-5197 (not to be published)

Jennifer Toren is a cybersecurity professional who has experience building and managing cybersecurity programs in the energy and aviation industries. She has a Master's in Technology Management from Georgetown University and has lived and worked in four countries: Canada, France, India, and the United States. She is passionate about integrating cybersecurity into business and IT processes to make security as easy as possible. Outside the office, she loves exploring new places, particularly by bicycle. She is also a member of Cloud Security Alliance's AI Organization Responsibility Working Group.
    Email: jennifer.toren@protonmail.com
    Phone: 703-967-1966

Chapter 9
Web3 and Quantum Attacks

Jerry Huang (ID) **and Ken Huang** (ID)

Abstract This chapter explores the intersection of quantum computing and Web3 technologies, focusing on both the challenges and solutions posed by the advent of quantum computing. The document begins with an examination of quantum mechanics principles as they apply to computing, delving into the current state and future projections of quantum computers. It then shifts focus to the potential threats quantum computing poses to Web3 and blockchain technologies, particularly highlighting vulnerabilities in cryptographic algorithms and outlining possible scenarios of quantum attacks. The broader implications of these threats on the Web3 ecosystem are also discussed. The latter part of the document is dedicated to exploring strategies for defending Web3 from quantum threats. This includes an analysis of quantum-resistant algorithms such as lattice-, hash-, and code-based cryptography, as well as the National Institute of Standards and Technology's (NIST) efforts in developing quantum-resistant algorithms. The integration of these algorithms into blockchain technology is scrutinized. Additionally, Quantum Key Distribution (QKD) and hybrid cryptographic solutions are presented as alternative defense mechanisms. A series of case studies provide practical insights into how these theoretical concepts are being applied in real-world scenarios.

Quantum attacks against blockchain technology are a real and present danger. The security of blockchains rests on the assumption that it would be impossible to crack the cryptographic algorithms used to create and verify blocks, but this may not be the case if quantum computers become available.

A quantum computer is a computer that uses quantum mechanical phenomena, such as superposition and entanglement, to perform operations on data. This means that a quantum computer can solve certain problems much faster than a classical computer.

J. Huang
The University of Chicago, Chicago, IL, USA
e-mail: jerryh@uchicago.edu

K. Huang (✉)
DistributedApps LLC, Fairfax, Virginia, USA
e-mail: ken@distributedapps.ai

© The Author(s), under exclusive license to Springer Nature
Switzerland AG 2024
K. Huang et al. (eds.), *Web3 Applications Security and New Security Landscape*,
Future of Business and Finance, https://doi.org/10.1007/978-3-031-58002-4_9

One problem that a quantum computer could solve is breaking the cryptographic algorithms used in blockchains. If a quantum computer could break these algorithms, it would be able to forge blocks and tamper with the blockchain. This could cause a lot of havoc in the blockchain world. Fortunately, there are ways to protect blockchains from quantum attacks. One way is to use quantum-resistant algorithms, which are algorithms that are not vulnerable to attacks from quantum computers. Another way is to use quantum key distribution (QKD), which is a way of securely exchanging cryptographic keys using quantum mechanics.

As we stand on the precipice of a quantum computing revolution, the very foundation of blockchain technology faces an unprecedented threat. Quantum computers, harnessing quantum mechanical phenomena, possess the potential to break cryptographic algorithms underpinning blockchains. While the immediate security of blockchains remains intact, the evolution of quantum computing necessitates proactive measures. This chapter unravels the implications of quantum attacks on Web3, delving into the mechanics of potential breaches and exploring the defenses available, such as quantum-resistant algorithms and QKD.

9.1 Quantum Computing: An Overview

To comprehend the threat quantum computing poses to Web3, it is imperative to first understand its mechanics. This section offers insights into quantum computing, contrasting its capabilities with classical computers and outlining its potential advantages.

9.1.1 Quantum Mechanics in Computing

The heart of quantum computing lies in quantum mechanics. This subsection elucidates quantum mechanical phenomena like superposition and entanglement, explaining their roles in enhancing computational prowess.

The realm of quantum computing is deeply rooted in quantum mechanics, a subfield of physics that explores the behavior of matter and energy at the smallest scales. Two quantum phenomena—superposition and entanglement—stand out as the essential building blocks that contribute to the extraordinary computational capabilities of quantum computers (QuTech, 2022).

At the core of quantum computing is the concept of the quantum bit, or qubit. Unlike classical bits in traditional computing, which can exist in one of two states— either 0 or 1—qubits can exist in a combination of these states simultaneously. This characteristic is due to the phenomenon of superposition. The ability of a qubit to be in multiple states at once creates the potential for parallel processing, which exponentially increases computational speed and efficiency. While a classical computer would have to check each entry in a database one by one to find a specific item, a quantum computer could, in theory, process all entries simultaneously, thus accelerating tasks like database searching, optimization problems, and even certain types of simulations.

Entanglement, another quantum phenomenon, takes the capabilities of quantum computing even further. When two qubits are entangled, the state of one qubit will instantaneously influence the state of another, no matter the distance that separates them. This interconnectedness allows quantum computers to perform complex calculations more efficiently than classical computers. Entanglement is leveraged in quantum algorithms to solve intricate problems, such as large-number factorization, at speeds unattainable by classical algorithms. The implications of this are significant, especially for cybersecurity, as it poses a potential threat to existing encryption methods.

How does the unique property of quantum entanglement contribute to the accelerated speed of quantum computing? Consider the following analogy: Picture a pair of dice rolled simultaneously. In classical mechanics, each die's outcome is independent of the other. However, in an entangled quantum state, the dice consistently display complementary results, regardless of the distance between their rolls.

Applying this concept to qubits, entangled qubits exhibit a phenomenon where the state of one qubit instantaneously influences the state of the other. This interconnection empowers quantum computers to manipulate multiple qubits simultaneously, resulting in a substantial increase in computational power.

Moreover, entanglement facilitates a high level of synchronicity. Operations performed on one qubit can instantly manifest corresponding results in its entangled partner. This instantaneous reflection significantly reduces computation time, a crucial aspect for addressing complex problems and handling large datasets.

Importantly, entanglement enables quantum computers to solve specific problems much more efficiently than classical computers. Especially in cryptography, where the factorization of large numbers is pivotal, quantum computers leverage entanglement to identify prime factors exponentially faster than their classical counterparts (more on this topic in Sect. 9.2). And this is the security threat we are talking about since modern web applications including web3 applications use cryptography which rely on computation complexity or factorization of large numbers (Thummalapenta, 2023).

Additionally, entanglement serves as the foundation for quantum teleportation, a process through which information is transferred from one location to another without physically traversing the space in between. Despite sounding like science fiction, quantum teleportation is a proven phenomenon that could revolutionize the speed of information transfer in quantum computing.

Understanding the principles of quantum computing is becoming increasingly important, especially for developers, architects, cybersecurity professionals, and college students. The reason is twofold: first, the computational power of quantum computers threatens to make current encryption techniques obsolete, requiring new, quantum-resistant methods. Second, the advanced capabilities of quantum computing open up new avenues for scientific research and data analysis, which could revolutionize industries ranging from healthcare to finance.

For those interested in integrating quantum computing into their existing technological infrastructure, numerous companies offer cloud-based quantum computing services. The process generally involves setting up a quantum computing environment using a preferred programming language, translating the problem at hand into a quantum algorithm, and finally executing the algorithm on a quantum processor. While this might sound straightforward, the real-world application involves a steep learning curve. The algorithms are complex and the interpretation of results is not always intuitive.

So, how do these principles of quantum mechanics translate into real-world computing applications? Consider the field of drug discovery. The interaction of molecules is essentially a quantum problem. Simulating these interactions with classical computers is time consuming and often inaccurate. However, quantum computers have the potential to model molecular structures much more efficiently and accurately (McKinsey, 2021). Similarly, in finance, optimizing portfolios is a computationally intensive task that could be streamlined with the use of quantum algorithms (Amazon Quantum Solutions Lab, 2023).

The principles of quantum mechanics not only form the foundation for quantum computing but also give it a computational advantage that classical computing can never achieve. While the field is still in its infancy, the potential applications are vast and the implications are profound, especially in cybersecurity and data analysis. As we stand on the cusp of a quantum revolution, understanding the quantum mechanical phenomena that power these advanced machines is crucial for anyone involved in technology and cybersecurity.

9.1.2 Quantum Computers: Current State and Future Projections

While quantum computing holds promise, where are we today? This subsection evaluates the current state of quantum computing, highlighting advancements and speculating on its future trajectory.

The landscape of quantum computing is a fascinating blend of theoretical promise and practical challenges. The enormous potential of quantum computers, as underscored by their theoretical speedup capabilities for certain problem classes, has garnered attention from academia, industry, and governments alike. However, the field is still largely experimental and faces numerous obstacles, both technical and conceptual, that must be overcome to realize its full potential. At the time of this writing which is December, 2023, the current state and future projections of quantum computing paint a complex but promising picture.

In terms of hardware, quantum computers have made significant strides but are still in a nascent stage. Companies like IBM, Google, and Rigetti are leading the charge in the development of quantum processors (Rogucki, 2023). For example, in December 2023, IBM unveiled its next-generation quantum processor, the IBM Quantum System Two, and has extended its roadmap to advance the era of quantum utility. The new system is designed to enable more powerful quantum computations and marks a significant step forward in the development of practical quantum applications. The company aims to continue driving progress in quantum computing to make it more accessible and useful for a variety of industries and scientific fields (IBM, 2023).

However, these processors are often limited by their "quantum coherence"—the time during which quantum information can be reliably stored and manipulated (Rouse, 2019). This time is usually extremely short, on the order of microseconds, which constrains the types of computations that can be performed. Extending this period of coherence is one of the central challenges in quantum hardware development. Additionally, the current quantum systems are "noisy," meaning they are susceptible to errors due to their interactions with the environment (Swayne, 2023). Error correction in quantum systems is a significant area of research, but practical, fault-tolerant quantum computers remain an aspirational goal for the future.

The development of quantum algorithms is another critical area that has seen notable progress but still has a long way to go. While algorithms like Shor's for factorization (Quantiki, 2015) and Grover's for database searching (Microsoft, 2023) demonstrate the theoretical superiority of quantum computing for certain problems, their practical implementation is often hindered by the limitations of current hardware. In essence, the algorithms are ahead of the hardware capabilities, awaiting the day when they can be effectively executed on stable, large-scale quantum systems.

The software ecosystem for quantum computing is comparatively more mature, thanks to high-level programming languages and platforms like Qiskit, Quipper, and Cirq (GQI, 2023). These platforms enable researchers and developers to experiment with quantum algorithms and simulate quantum computations, even if they do not have access to a physical quantum computer. This has facilitated a broader engagement with quantum computing from the software development and data science communities, who can now experiment with quantum algorithms to solve practical problems, albeit in a simulated environment.

In terms of applications, while full-scale, practical quantum computers are not yet available, there are areas where "quantum advantage" can be achieved using existing hardware for specific problems. For instance, quantum annealers like D-Wave systems have been used in optimization tasks for logistics and supply chain management (Swayne, 2021). Though these are not universal quantum computers, they do offer a glimpse into the kinds of real-world problems that quantum computing could eventually solve more efficiently.

Looking forward, the trajectory of quantum computing is likely to be characterized by incremental advancements rather than sudden breakthroughs. We can expect gradual improvements in quantum coherence times and error rates, accompanied by the development of more efficient algorithms and fault-tolerant techniques. Public and private investments in quantum research are likely to increase, given the strategic importance of quantum computing in national security, economic competitiveness, and scientific advancement.

Moreover, as quantum computing matures, we can anticipate a paradigm shift in several fields, including but not limited to, cryptography, machine learning, and material science. The advent of practical quantum computers will necessitate entirely new cryptographic techniques, accelerate data analysis algorithms, and enable highly accurate simulations of molecular structures.

A recent report highlights the imminent threat posed by quantum computers to current encryption standards, a concern voiced by Tilo Kunz of Canadian cybersecurity firm QD5 (LAGUE, 2023). Predicting that by 2025, quantum computers could render existing digital security measures obsolete, Kunz emphasizes the urgency of the situation, especially for sensitive data like military secrets and personal health records. Major powers, particularly the USA and China, are in a race to develop these quantum capabilities while also accusing each other of large-scale data harvesting. The report underscores the transformative potential of quantum technology. Amidst this technological arms race, efforts are underway to develop new encryption methods, known as post-quantum cryptography, to safeguard against the quantum threat. However, there is a divide among experts on the effectiveness and timeline of quantum computing's impact, with some predicting its significant influence only by mid-century. As the world anticipates "Q-day," the point at which quantum computers can break current encryption, nations and companies are bracing for a new era of digital security challenges.

To sum up, the current state of quantum computing is one of cautious optimism. While substantial hurdles remain, the field has made impressive strides in both theoretical and practical aspects. The future holds the promise of revolutionizing our computational capabilities, albeit with challenges that will require concerted efforts from researchers, engineers, and policymakers alike. For professionals involved in technology and cybersecurity, keeping abreast of developments in quantum computing is not merely optional; it is imperative for understanding the evolving landscape of computational possibilities and vulnerabilities.

9.2 Quantum Threats to Web3 and Blockchain

The intersection of quantum computing and blockchain poses both potential and peril. This section dives deep into the vulnerabilities that quantum computers could exploit within blockchain technology and Web3.

9.2.1 Cryptographic Algorithm Vulnerabilities

The bedrock of blockchain's security is its cryptographic algorithms. However, quantum computers could destabilize this foundation. This subsection explores the cryptographic algorithms at risk and the potential aftermath of their breaches.

Blockchain technology has gained some level of adoption due to its robust security features, which are fundamentally based on cryptographic algorithms. These algorithms ensure data integrity, user anonymity, and secure transactions within a decentralized network. However, the advent of quantum computing poses a looming threat to the cryptographic foundation on which blockchain relies. Specifically, quantum computers have the potential to undermine the cryptographic algorithms that have long been considered secure by classical standards. This subsection aims to delve into the types of cryptographic algorithms at risk and explore the consequences if these cryptographic bedrocks were to be compromised by quantum capabilities.

Firstly, let us discuss the types of cryptographic algorithms most at risk. Hash functions, such as SHA-256 used in Bitcoin, are considered relatively more resilient to quantum attacks. However, public-key cryptographic algorithms, which form the basis for digital signatures and secure key exchange, are particularly vulnerable. Algorithms like RSA and Elliptic Curve Cryptography (ECC) are often used in blockchain for generating digital signatures and securing transactions. These algorithms fundamentally rely on the computational intractability of certain mathematical problems, like integer factorization in the case of RSA and the elliptic curve discrete logarithm problem for ECC. Quantum algorithms, such as Shor's algorithm (see Box Shor's algorithm), can solve these problems exponentially faster than the best-known classical algorithms. As a result, if and when scalable, fault-tolerant quantum computers become a reality, they could potentially decrypt secure keys and forge digital signatures, thereby compromising the integrity and security of the entire blockchain network.

Shor's Algorithm

Shor's Algorithm is a quantum algorithm (Pavlidis & Gizopoulos, 2022) for efficiently factoring large numbers, which is significantly faster than any known classical approach.

1. Problem Addressed: Factoring a large composite number (N) into its prime factors, a task that is computationally challenging for classical computers.
2. Quantum Mechanics Utilization: Uses quantum mechanics principles like superposition (handling multiple states simultaneously) and entanglement (correlating the states of particles).
3. Steps Involved:

    - Quantum Fourier Transform (QFT): Transforms the state of a quantum system to reveal periodicity (Mastriani, 2021).
    - Finding Periodicity: Identifies the period (r) of the function ($f(x) = a^x$ mod N), where (a) is a random number less than (N). This is done using quantum superposition.
    - Measurement and Classical Computation: After applying QFT and measuring the quantum state, classical methods are used to derive factors of (N) from the period (r).

4. Efficiency: Operates in polynomial time relative to the number of digits in (N), an exponential improvement over classical algorithms.
5. Implications: Challenges the security of cryptographic systems like RSA and ECC (SSL2BUY, 2022), which depend on the difficulty of factorizing large numbers.

This algorithm exemplifies the potential of quantum computing to solve specific problems much more efficiently than traditional computers.

The implications of such a breach would be manifold and catastrophic. Firstly, the compromise of digital signatures would mean that malicious actors could forge transactions, essentially enabling them to double-spend or even steal digital assets. This could lead to a complete loss of trust in the blockchain network in question, causing the value of the associated digital currency to plummet. Secondly, if secure key exchange mechanisms are broken, it opens up the possibility of eavesdropping on encrypted communications between nodes in the blockchain network. This could compromise not only financial transactions but also any confidential data stored in a blockchain-based system.

Further, beyond financial losses, such a breach could have regulatory implications. Many blockchain networks are being explored for use-cases in legal (De Miguel, 2023), healthcare (Vijayakumar, 2022), and governmental systems (Brothwell, 2023). A compromise of cryptographic security in these contexts could mean unauthorized access to sensitive information, ranging from personal identification data to state secrets, which could have national security implications. In a worst-case scenario, this could lead to a large-scale reconsideration and rollback of blockchain implementations across multiple sectors, dealing a severe blow to the advancement of decentralized systems.

The cybersecurity community is not unaware of these risks. There is active research in the field of post-quantum cryptography, aiming to develop new cryptographic algorithms that would be secure against the quantum threat. Lattice-, hash-, and code-based cryptography are some of the avenues being explored for this purpose (see Box below). The idea is to transition to these quantum-resistant algorithms as the field of quantum computing continues to advance. However, this transition would be neither simple nor instantaneous, given the widespread and decentralized nature of blockchain networks. It would require a coordinated effort from developers, users, and possibly even regulatory bodies to successfully migrate to new cryptographic standards without compromising the integrity of existing systems.

Some Post-Quantum Cryptography Algorithms
Lattice-, hash-, and code-based cryptography are key types of post-quantum cryptographic algorithms. These methodologies are being developed to secure digital communications against the potential threat posed by quantum computers, which are expected to have the capability to break many of the current cryptographic systems.

1. Lattice-Based Cryptography (Ahmad, 2022): This type of cryptography relies on the mathematical complexity of lattice problems, which involve multidimensional point grids. It is considered hard for quantum computers to solve due to the complexity of finding short vectors in these high-dimensional lattices. Lattice-based cryptographic systems are promising for their efficiency and the relative ease of implementing key operations

like encryption and decryption. Notable examples include the Learning With Errors (LWE) and the Shortest Vector Problem (SVP). LWE is a cryptographic concept based on the difficulty of solving mathematical problems that have been intentionally distorted with small random errors. It is considered secure against quantum computer attacks because correcting these errors to find the original solution is extremely challenging, even for quantum computers. LWE forms the foundation for many post-quantum cryptographic systems. SVP in cryptography is a mathematical challenge involving finding the shortest, non-zero vector in a lattice. This problem is computationally difficult to solve, especially in higher dimensions, making it a robust foundation for certain cryptographic systems, particularly in the realm of post-quantum cryptography.

2. Hash-Based Cryptography (de Quehen, 2020): This method employs cryptographic hash functions, which are mathematical algorithms that convert input data of any size into a fixed-size string of characters. Hash-based cryptography is mainly used for creating digital signatures, ensuring data integrity, and authenticating information. It is believed to be resilient against quantum computing attacks because these attacks do not significantly improve upon the best-known classical attacks against hash functions. Merkle trees are a common structure used in hash-based cryptography.

3. Code-Based Cryptography (Cook, 2019): Originating from error-correcting codes, code-based cryptography is another alternative resilient to quantum attacks. It involves encoding messages in a way that intentionally introduces errors, which only the legitimate receiver can correct using a secret decoding key. The security of code-based cryptography hinges on the difficulty of decoding a general linear code, a problem that is not expected to be efficiently solvable by quantum computers. The most famous example is the McEliece cryptosystem.

Each of these post-quantum cryptographic approaches offers a unique set of advantages and challenges. The selection of which method to use often depends on specific requirements like computational efficiency, key size, and the nature of the data being protected. As quantum computing continues to advance, the development and implementation of these post-quantum cryptographic methods are becoming increasingly important to ensure the security of digital communications and data. For more details on these algorithms, please see Sect. 9.3.1.

In essence, while blockchain technology offers a robust and secure framework for a variety of applications, its reliance on cryptographic algorithms that are vulnerable to quantum attacks is a significant concern. The potential compromise of these algorithms would have far-reaching implications, from financial and data losses to regulatory and national security risks. The challenge for the cybersecurity community is twofold: to accelerate the development and adoption of quantum-resistant cryptographic algorithms and to prepare for a seamless transition to these new algorithms to secure existing and future blockchain networks. Given the gravity and complexity of these challenges, it is imperative for professionals in technology and cybersecurity to stay abreast of both advancements in quantum computing and developments in post-quantum cryptography.

9.2.2   Quantum Attacks: Possible Scenarios

How might a quantum attack on a blockchain unfold? This subsection paints possible scenarios of quantum breaches. Figure 9.1 gives a high-level overview of these attacks.

A quantum attack on a blockchain would represent a watershed moment in both the fields of cybersecurity and quantum computing. The potential for quantum computers to break cryptographic algorithms forms the basis for such an attack. While full-scale quantum computers capable of carrying out these attacks do not yet exist, it is instructive to envision possible scenarios to better understand the risks and prepare for future contingencies. In this subsection, we will explore hypothetical but plausible scenarios in which quantum attacks could unfold, focusing on how attackers might forge blocks and disrupt the integrity of blockchains.

One conceivable scenario begins with the compromise of digital signatures. In traditional blockchain networks like Bitcoin, transactions are verified through digital signatures based on Elliptic Curve Cryptography (ECC, see Box ECC). An attacker with a sufficiently powerful quantum computer could employ algorithms like Shor's algorithm to break this cryptography, essentially giving them the ability to forge digital signatures. Once the digital signatures are compromised, the attacker could make unauthorized transactions appear legitimate. For example, they could transfer funds from other users' wallets to their own, essentially stealing digital assets. Since the transactions would bear a valid (though forged) digital signature, the network would be none the wiser and would confirm the transactions, adding them to the blockchain.

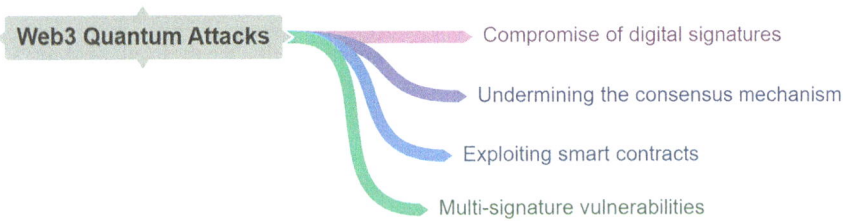

Fig. 9.1 Web3 quantum attacks

ECC

Elliptic Curve Cryptography (ECC) is a method of public-key cryptography based on the algebraic structure of elliptic curves over finite fields. ECC is used in various digital security applications, including blockchain technology, due to its efficiency and high level of security.

In the context of blockchain, ECC plays an important role in ensuring the security and integrity of transactions. Here is how ECC is typically utilized in blockchain systems:

1. Key Generation: ECC is used to create public and private key pairs. In blockchain, each participant (or node) has a unique pair of keys. The private key, which is kept secret, is used to sign transactions and prove ownership of a blockchain address. The public key, derived from the private key, is shared openly and acts as an address to which others can send transactions.
2. Transaction Signing: When a user initiates a transaction, such as transferring cryptocurrency, they sign the transaction with their private key. This signature, created using ECC algorithms, proves that the transaction was initiated by the rightful owner of the funds without revealing the private key itself.
3. Transaction Verification: Other nodes in the blockchain network use the sender's public key to verify the signature. If the signature is valid, it confirms that the transaction was indeed created by the owner of the private key, ensuring that the transaction is legitimate and has not been tampered with.
4. Security: ECC provides a high level of security even with relatively small key sizes, making it efficient for use in blockchain networks where speed and resource efficiency are important. For example, a 256-bit key in ECC is considered to provide comparable security to a 3072-bit key in RSA (another public-key cryptography system).
5. Smart Contracts and DApps: In the development of decentralized applications (DApps) and smart contracts, ECC can be used to manage permissions and create secure connections between parties. This aspect is particularly relevant in environments where trustless interactions are required.

ECC's effectiveness in blockchain is due to its ability to offer a high level of cryptographic security with lower computational overhead compared to other methods like RSA. This efficiency is particularly beneficial in blockchain networks, where performance and scalability are key concerns. The use of ECC in blockchain technology thus contributes significantly to the secure and efficient functioning of decentralized digital ledgers. As stated previously in this chapter, while highly secure against classical computing attacks, ECC is considered susceptible to quantum computing due to the nature of quantum algorithms.

computational power. In such a situation, the attacker could manipulate the block-chain to double-spend coins and prevent other transactions from getting confirmed, disrupting the network's functionality and eroding trust in the system. Scientists have recently proposed a quantum PoW consensus algorithm in an effort to deal with this potential problem (Greene, 2023). This new approach to PoW is claimed to offer several advantages over traditional methods. Firstly, it promises increased security through quantum-resistance, defending against potential attacks from quantum computers. Secondly, validation of transactions is hailed to see dramatic speedup compared to what is achievable with classical PoW. Thirdly, energy consumption is touted to decrease substantially relative to existing PoW protocols. However, the proposal remains in the research phase for now. While initial findings appear promising, further testing and development are needed to ascertain its real-world feasibility and bring the concept to fruition. At the current stage, the capabilities are so far theoretical and need to undergo more rigorous evaluation before any implementations can be rolled out.

Furthermore, a quantum attacker could exploit smart contracts on platforms like Ethereum. Smart contracts are self-executing contracts with coded rules, and they often rely on the same vulnerable cryptographic functions as transactions. By breaking these, an attacker could, for instance, alter the conditions of a smart contract to divert funds to their account, trigger unauthorized actions, or even freeze the contract altogether.

It is worth mentioning that multi-signature wallets, often used for added security by requiring multiple parties to approve a transaction, would not offer protection against a quantum attack. Multi-signature algorithms are generally as quantum-vulnerable as single-signature ones; therefore, an attacker with a quantum computer could forge multiple signatures as easily as a single one.

The aftermath of such quantum attacks would be devastating. Beyond the immediate financial losses, the integrity of the affected blockchain would be irrevocably damaged. Restoring confidence in the system would be a monumental challenge, necessitating an urgent transition to quantum-resistant cryptographic algorithms. This would be a complicated, resource-intensive process, requiring coordinated efforts from network participants and possibly oversight from regulatory bodies.

Please note that these scenarios are not immediate threats but rather cautionary tales for what could happen if quantum computing progresses faster than the adoption of quantum-resistant security measures. As such, they serve as a call to action for researchers, cybersecurity professionals, and policymakers to accelerate advancements in post-quantum cryptography and to begin planning for a future where quantum attacks are not just possible but probable. The development of comprehensive defense strategies, including real-time monitoring for unusual network activities and rapid response plans for potential breaches, will be essential components of a multi-layered defense strategy against the quantum threat. While we cannot predict with certainty when quantum computers will become powerful enough to execute such attacks, preparing for these scenarios is not just prudent; it is imperative for the long-term viability and integrity of blockchain networks.

9.2.3 Broader Implications to Web3 Ecosystem

Beyond blockchains, the Web3 ecosystem comprises various applications and systems. This section delves into the broader implications of quantum attacks on the expansive Web3 landscape.

The Web3 ecosystem, often dubbed as the new paradigm for applications on the Internet, aims to create a decentralized and distributed network that goes beyond just blockchains. It encompasses various components like decentralized finance (DeFi), decentralized autonomous organizations (DAOs), non-fungible tokens (NFTs), Metaverse applications and even decentralized Internet protocols. Given its wide scope, a quantum attack would have ramifications that extend well beyond just compromising a single blockchain. This section aims to explore the broader implications such attacks would have on the expansive Web3 landscape.

First, let us consider decentralized finance or DeFi, which has become a cornerstone of the Web3 ecosystem. DeFi platforms often use smart contracts to facilitate financial transactions without the need for traditional intermediaries like banks. These smart contracts, like the blockchains they run on, rely on cryptographic algorithms for security. A quantum attack that compromises these algorithms could enable unauthorized transactions, asset theft, or even the alteration of the contract terms. Imagine a lending protocol where an attacker could manipulate interest rates or a decentralized exchange where asset prices could be artificially manipulated; the entire financial structure of DeFi could be destabilized.

Next, consider non-fungible tokens (NFTs), which have gained significant attention for their ability to provide provable ownership of digital assets. The ownership and transfer of NFTs are facilitated by cryptographic algorithms. If these were to be broken, it would be possible to forge ownership, effectively making the unique, "non-fungible" tokens entirely fungible and worthless. The collapse of trust in the NFT market could have ripple effects on various industries, including art, gaming, and even real estate, where NFTs are beginning to find applications.

Decentralized Autonomous Organizations (DAOs) are another crucial component of the Web3 landscape. These are entities that operate without centralized control, governed by smart contracts and consensus among its members. The integrity of the voting process in a DAO, often used for decision-making, relies on cryptographic algorithms. A quantum attack could compromise the voting mechanism, allowing an attacker to hijack the DAO and make unauthorized decisions, such as misappropriating funds or altering the DAO's rules.

Beyond financial systems and organizations, even the lower-level protocols of the Web3 ecosystem could be vulnerable. Technologies like the InterPlanetary File System (IPFS) and the Solidity programming language, which are foundational to the decentralized web, employ cryptographic algorithms for data integrity and access control. The compromise of these algorithms would have a cascading effect, jeopardizing not just individual applications but the entire architecture of the Web3 ecosystem.

The broader implications of a quantum attack on the Web3 ecosystem would be a mix of immediate chaos and long-term distrust. Financial losses could run into billions, and the social and psychological impact of widespread fraud and the collapse of what was considered a secure and decentralized future would be severe. Legal repercussions could also be significant, with potential lawsuits and even government intervention calling the whole idea of decentralization into question.

To mitigate these risks, the transition to quantum-resistant algorithms is not just advisable but imperative. This is a complex undertaking that will require concerted efforts from developers, users, and perhaps even regulatory bodies. Aside from cryptographic changes, Web3 platforms will need to incorporate real-time monitoring systems capable of identifying and flagging unusual activities that could indicate a quantum attack. Contingency plans, including rapid response mechanisms and data backup protocols, will also need to be developed.

In conclusion, the advent of quantum computing is a double-edged sword for the Web3 ecosystem. While it brings the promise of unmatched computational capabilities, it also poses a significant threat to the cryptographic foundations of various Web3 components. Preparing for the potential risks associated with quantum attacks is not just a technical necessity but a critical step in ensuring the long-term viability and trustworthiness of the Web3 landscape. The task is monumental but ignoring it could result in consequences far more severe than any challenge posed by the transition. For anyone involved in the development or utilization of Web3 technologies, understanding and preparing for the implications of quantum computing should be considered a top priority.

9.3 Defending Web3 from Quantum Threats

All is not bleak in the face of quantum threats. Innovations in cryptography and quantum mechanics offer robust defense mechanisms. This section elucidates these defenses, from quantum-resistant algorithms to the pioneering QKD. Figure 9.2 provides an outline of this section.

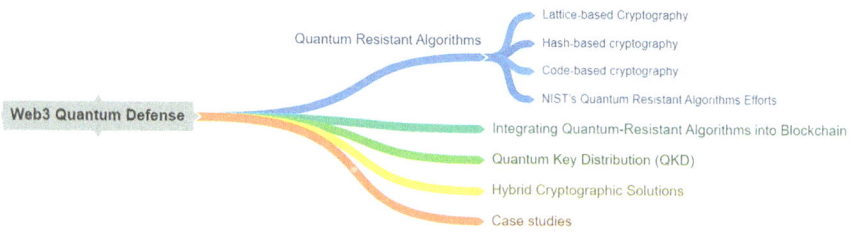

Fig. 9.2 Defending Web3 from quantum threats

9.3.1 Quantum-Resistant Algorithms

The first line of defense against quantum threats is the development and adoption of quantum-resistant algorithms. This subsection discusses these algorithms, their mechanics, and their integration into blockchains.

The looming threat of quantum computing on cryptographic systems has accelerated the quest for quantum-resistant algorithms. These algorithms are designed to be secure against the capabilities of quantum computers, offering a robust line of defense against potential quantum attacks. This subsection will discuss the nature of quantum-resistant algorithms, delve into the mechanics that make them secure, and explore how they can be integrated into existing blockchain systems as a precautionary measure against quantum vulnerabilities.

Quantum-resistant algorithms often rely on mathematical problems that, as of current understanding, do not have an efficient quantum solution. While algorithms like RSA and ECC are vulnerable to Shor's algorithm, quantum-resistant algorithms leverage problems like lattice-based cryptography, hash-based cryptography, multivariate polynomial cryptography, and code-based cryptography. Each of these mathematical frameworks provides a different approach to securing data, making it computationally infeasible for even a quantum computer to crack within a reasonable time frame.

We have discussed these algorithms briefly in Sect. 9.2.1. Let us delve a little deeper into some of these types of quantum-resistant cryptography to understand their mechanics.

9.3.1.1 Lattice-Based Cryptography

Lattice-based cryptography relies on the mathematical concept of lattices in multidimensional space. Lattices are essentially grids of points that extend infinitely in every direction within these spaces. The appeal of lattice-based cryptography, especially in today's context, is its potential resistance to quantum computing attacks, making it a strong candidate for securing communications in a future where quantum computers are commonplace.

To visualize a lattice, imagine a regular, infinite grid in a multidimensional space. Each point on this grid represents a position that can be reached by stepping in various directions and distances from a starting point, using a set of fixed rules. In cryptography, these points and the rules for reaching them form the basis of creating and deciphering secure messages.

Lattice-based cryptography's security is based on the difficulty of solving certain problems:

1. Shortest Vector Problem (SVP): This involves finding the shortest path from the origin point (zero) to any other point in the lattice. As you can imagine, in a high-dimensional space with a near-infinite number of points, finding the absolute shortest path is extremely challenging.

2. Closest Vector Problem (CVP): Here, the challenge is to find the nearest lattice point to a given non-lattice point. Again, in a complex, multidimensional lattice, this task becomes highly difficult.
3. Learning With Errors (LWE): This problem involves working with pairs of data where one element of the pair is a random piece of data and the other is a mixture of a secret piece of data and some error or noise. The goal is to figure out the secret data from many such pairs, which is akin to finding a pattern in a very noisy environment.

Lattice-based cryptography is used to develop:

1. Encryption methods: These methods encode messages as points in a lattice. The security comes from the added complexity of small errors, which makes decoding the message correctly almost impossible without the specific key.
2. Digital Signatures: Similar to traditional digital signatures but based on lattice problems, these signatures ensure a message's authenticity and the sender's identity, with the added benefit of being secure against quantum attacks.
3. Key Exchange Protocols: These protocols enable two parties to agree on a secret key over an insecure channel. They use lattices to ensure that even if someone is listening, they cannot figure out the agreed-upon key.

Lattice-based cryptography advantages include quantum resistance, making it a promising solution for future-proofing secure communications against quantum computing threats. Additionally, many lattice-based algorithms are efficient in terms of computation and data size.

However, there are challenges such as the complexity in designing and implementing these systems, and the ongoing process of standardization and evaluation compared to more established cryptographic methods.

9.3.1.2   Hash-Based Cryptography

Hash-based cryptography is a vital branch of cryptography that employs hash functions to transform input data of any size into a fixed-size string of characters. These mathematical algorithms are designed to make this transformation a one-way process, rendering it computationally infeasible to reverse-engineer the original input from the hash output. The essence of hash functions lies in several key characteristics: they are deterministic, meaning the same input consistently produces the same output; they are designed for quick computation; they possess pre-image resistance, making it extremely difficult to deduce the original input from its hash; they demonstrate an avalanche effect, where small changes in input lead to significant changes in output; and they exhibit collision resistance, greatly reducing the probability of two different inputs yielding the same hash output.

In the realm of digital signatures, hash-based cryptography finds its most prominent application. The process involves computing the hash of a message and encrypting it with a private key to generate a signature. This signature is then verified by applying the same hash function to the original message; if the decrypted signature matches the computed hash, the signature is considered valid. This method also plays a crucial role in ensuring data integrity, as comparing the hash values of the original and received data can indicate whether the data was altered during transmission. Furthermore, hash functions are commonly used for secure password storage, storing only the hash of the password rather than the password itself, thereby enhancing security during authentication processes.

Another critical application of hash functions is in the construction of Merkle trees in blockchain technology, facilitating efficient and secure verification of large data structures, such as transaction blocks. Hash-based signature schemes, such as Lamport signatures and Merkle signature schemes, are recognized for their potential in post-quantum cryptography. These schemes rely on the security of hash functions and are designed to be resistant to quantum computer attacks, as they do not depend on the difficulty of mathematical problems that quantum computers can solve efficiently.

While hash-based cryptography offers simplicity and efficiency, and its signature schemes are resistant to quantum computing attacks, there are challenges to be addressed. For instance, in hash-based signature schemes, the sizes of keys and signatures can be quite large, posing drawbacks in terms of storage and transmission efficiency. Additionally, many of these signature schemes, like the Lamport signature, are one-time signatures, meaning they can securely sign only one message with a given set of keys. Despite these challenges, hash-based cryptography remains a cornerstone in the field of digital security, with its applications in data integrity, secure password storage, digital signatures, and blockchain technology underscoring its versatility and significance. As we edge closer to the era of quantum computing, the role of hash-based cryptography in developing quantum-resistant cryptographic systems becomes increasingly pivotal.

9.3.1.3   Code-Based Cryptography

Code-based cryptography represents a unique approach within the cryptographic domain, utilizing principles of error-correcting codes from information theory. This type of cryptography gains its significance from its anticipated resistance to quantum computer attacks, thereby positioning itself as an integral part of the future of post-quantum cryptography. The essence of code-based cryptography lies in the challenge of decoding a general linear code, a task known to be computationally intensive.

At the heart of code-based cryptography is the use of linear error-correcting codes. These codes are traditionally employed to safeguard data against errors during transmission or storage. In the context of cryptography, the message is encoded using these codes, which introduces redundancy and thereby facilitates error correction. The unique twist in code-based cryptography is the deliberate introduction of errors in a controlled manner as part of the encryption process, making the task of unauthorized decoding exceedingly difficult.

A prominent example of code-based cryptography is the McEliece cryptosystem, established by Robert McEliece in 1978. In this system, the key generation process involves selecting a random, efficient-to-decode error-correcting code and keeping its decoding algorithm secret. The public key, derived from this code, is modified in such a way that decoding it without the secret key is as complex as solving the general problem of decoding an arbitrary linear code. During encryption, the sender encodes the message using the public key and intentionally adds a number of errors. The receiver, possessing the secret key, which includes knowledge of the original error-correcting code and its decoding algorithm, is able to decode the received message, correct the errors, and retrieve the original message.

The security of code-based systems primarily relies on the difficulty of decoding a random linear code, a problem that remains unsolved even in the realm of quantum computing. This resistance to quantum attacks is a key advantage, as it ensures the relevance of code-based cryptography in a future where quantum computing is prevalent. Additionally, these systems are often lauded for their efficiency in encryption and decryption processes, boasting relatively low computational complexity compared to other cryptographic methods.

However, code-based cryptography is not without its challenges. The most notable of these is the size of the keys involved; for instance, the public keys in the McEliece cryptosystem are substantially larger than those used in conventional systems like RSA or ECC. This size issue poses practical limitations in terms of storage and data transmission. Furthermore, as an area of ongoing research and development, efforts continue to optimize key sizes and enhance the overall efficiency of these systems, ensuring their robustness against both classical and quantum computing threats.

In essence, code-based cryptography offers a compelling solution for secure communications, especially in an era increasingly dominated by quantum computing advancements. Its foundation on the complex problem of decoding arbitrary linear codes provides a strong security basis, making it a notable and evolving field within the broader landscape of cryptography.

9.3.1.4 NIST's Quantum-Resistant Algorithms Efforts

The National Institute of Standards and Technology (NIST) is at the forefront of pioneering encryption algorithms capable of resisting attacks by quantum computers. Following the selection of four resilient algorithms in 2022 (see Box NIST Selected Algorithm), NIST has embarked on the final phase of standardizing these cryptographic tools. This advancement marks a significant milestone, as it prepares the global cryptographic community to integrate these algorithms into existing encryption infrastructures (NIST, 2023).

NIST Selected Algorithm

In 2022, the National Institute of Standards and Technology (NIST) selected four algorithms as part of its ongoing efforts to develop standards for post-quantum cryptography. These algorithms were chosen for their resilience against potential attacks from quantum computers. The four selected algorithms are:

1. CRYSTALS-Kyber: This algorithm is primarily designed for key encapsulation, a process crucial for securing encrypted communications. CRYSTALS-Kyber is based on the hardness of problems in structured lattices, which are mathematical constructs known for their complexity and resistance to quantum attacks.
2. CRYSTALS-Dilithium: Selected for digital signature purposes, CRYSTALS-Dilithium also relies on the hardness of problems in structured lattices. Digital signatures are essential for verifying the authenticity and integrity of digital messages and documents, making this algorithm vital in the realm of secure digital communications.
3. SPHINCS+: This is a state-of-the-art hash-based signature scheme. Unlike the other three algorithms, SPHINCS+ does not rely on the hardness of lattice problems but instead on the security of hash functions. Hash-based cryptography is known for its simplicity and efficiency, and it is also believed to be resistant to quantum computing threats.
4. FALCON (Fast-Fourier Lattice-based Compact Signatures over NTRU): FALCON is another algorithm chosen for digital signatures. It uses lattice-based cryptography, specifically leveraging the NTRU (Nth-degree truncated polynomial ring) problem, which is considered hard for both classical and quantum computers to solve. The use of Fast-Fourier Transform techniques in FALCON aims to make the algorithm more efficient and compact.

These algorithms represent a diverse set of cryptographic approaches, each addressing different aspects of encryption and digital signatures. Their selection by NIST is a significant step in the transition toward securing digital systems against the emerging threat of quantum computing.

Recently, NIST released draft standards for three of the initially selected algorithms. These drafts signify a critical step toward creating practical standards that organizations worldwide can utilize. The fourth algorithm, FALCON, is expected to have its draft standard published in about a year. The journey to these quantum-resistant algorithms began in 2016, with NIST's call to cryptographic experts for submissions. This resulted in 69 candidate algorithms submitted by experts worldwide, which underwent rigorous analysis and testing in a transparent and collaborative process. These evaluations aimed to reduce the number of candidates and identify the most robust algorithms for standardization.

The new publications, designated as draft Federal Information Processing Standards (FIPS), detail each of the selected algorithms:

- FIPS 203 covers CRYSTALS-Kyber, designed for general encryption purposes like creating secure websites (Baker, 2023a, b).
- FIPS 204 discusses CRYSTALS-Dilithium, intended for protecting digital signatures.
- FIPS 205 covers SPHINCS+ which is a Stateless Hash-Based Digital Signature Standard, used for creating digital signatures.

9.3.1.5 Integrating Quantum-Resistant Algorithms into Blockchain

The process of integrating these quantum-resistant algorithms into existing blockchain networks presents several challenges but is crucial for long-term security. The decentralized nature of blockchains makes network-wide upgrades cumbersome, as they would require consensus among all participants. Such an upgrade, often termed a "hard fork," could be divisive and lead to splits in the network, as seen in past instances like the Bitcoin and Ethereum forks. However, the gravity of the quantum threat might serve as a unifying factor, leading to more streamlined adoption of new cryptographic standards.

The integration process itself would involve multiple steps. First, a candidate quantum-resistant algorithm would need to be rigorously tested for both security and performance. Given that quantum-resistant algorithms often involve more complex computations, it is vital to ensure that the added security does not come at the cost of significantly reduced transaction speeds or increased computational requirements. Once a suitable algorithm is selected, it would need to be implemented in a test environment to assess its real-world performance metrics and potential vulnerabilities. Following successful testing, the algorithm could then be rolled out as part of a network-wide upgrade.

While the shift to quantum-resistant algorithms is a significant undertaking, it is not the only measure needed to secure a network against quantum threats. It should be part of a multi-layered security approach that also includes real-time monitoring for malicious activity, secure backup protocols, and a contingency plan for immediate action in the event of a suspected quantum attack. This comprehensive strategy ensures that even if one line of defense is compromised, multiple other layers provide additional security.

In summary, quantum-resistant algorithms represent a critical first line of defense against the vulnerabilities introduced by quantum computing. These algorithms are based on mathematical problems that remain secure even in the face of quantum computational capabilities. While integrating these algorithms into existing blockchain networks presents logistical and technical challenges, the severity of the quantum threat makes it an imperative task. The adoption of quantum-resistant algorithms should be viewed as a necessary evolution of blockchain technology, enabling it to withstand emerging threats and continue serving as a secure, decentralized foundation for a wide range of applications. Given the complexities involved, the transition to quantum-resistant cryptography is not just a task for developers but requires the active engagement of all stakeholders, including users, cybersecurity experts, and potentially even regulatory bodies, to ensure a smooth and secure transition into this new era of quantum-safe blockchain technology.

9.3.2   Quantum Key Distribution (QKD)

QKD offers a novel approach to exchanging cryptographic keys, harnessing the principles of quantum mechanics (Grammel et al., 2021). Here, we explore the concept of QKD, its advantages, and its potential role in safeguarding Web3 applications.

QKD stands as a paradigm shift in the realm of secure communications, turning the threat posed by quantum computing on its head by using quantum mechanics as a tool for enhanced security. The essence of QKD lies in the quantum behavior of particles like photons for the secure exchange of cryptographic keys between parties. Unlike classical methods of key distribution, which could be vulnerable to computational attacks, QKD offers a level of security based on the fundamental principles of quantum physics. In this subsection, we will examine the concept of QKD, its inherent advantages, and how it could be instrumental in securing Web3 applications and the broader decentralized ecosystem.

In a conventional cryptographic system, the security of a key exchange relies on the computational difficulty of solving certain mathematical problems. However, as we have discussed, quantum computers have the potential to solve these problems exponentially faster than classical computers, thereby jeopardizing the entire encryption scheme. QKD sidesteps this vulnerability by employing the quantum properties of particles. For example, in a typical QKD setup using polarized photons, the quantum states of the photons themselves serve as the cryptographic key. The laws of quantum mechanics dictate that any attempt to eavesdrop on this key would inevitably alter the quantum states of the particles, thus revealing the intrusion. This feature, known as quantum indeterminacy, ensures that the key exchange process is secure against any kind of computational attack, classical or quantum.

One of the primary advantages of QKD is its ability to provide unconditional security, rooted in the laws of physics rather than computational complexity. This means that, theoretically, no advancement in computational capabilities can compromise a QKD system. Moreover, QKD enables the detection of eavesdropping attempts in real time, providing an additional layer of security. If any interception attempt is detected, a new key can be generated instantaneously, rendering the intercepted key useless.

Integrating QKD into the Web3 ecosystem could offer a robust solution for secure communications in decentralized networks. Whether it is secure transactions in a blockchain, confidential communications in decentralized organizations, or private data storage in distributed file systems, QKD could provide an unbreakable encryption layer. For example, in a blockchain network that incorporates QKD, the transaction data could be encrypted with keys exchanged via quantum channels, making it virtually impossible for malicious actors to gain unauthorized access to the transaction details.

However, the adoption of QKD in Web3 applications is not without challenges. First and foremost, the technology requires specialized hardware capable of generating, transmitting, and detecting quantum states, which can be both expensive and difficult to deploy at scale. Additionally, the effective range for QKD is currently limited due to photon loss in optical fibers, although research into quantum repeaters aims to extend this range. Furthermore, integrating QKD into existing networks would require significant architectural changes, making it a long-term rather than an immediate solution.

Despite these challenges, the unparalleled security advantages of QKD make it a compelling option for safeguarding the future of Web3 applications. As quantum computing technologies mature, the urgency to transition to quantum-secure methods will only intensify. In this context, QKD represents not just an optional enhancement but a necessary evolution in cryptographic practices. While the implementation of QKD will demand substantial investment in infrastructure and research, the payoff in terms of security could be invaluable.

9.3.3   Hybrid Cryptographic Solutions

By combining classical and quantum-resistant algorithms, hybrid solutions aim to offer a layered defense against both classical and quantum threats. This approach ensures immediate security while also future-proofing Web3 systems against emerging quantum threats.

Hybrid cryptographic solutions represent a pragmatic approach to navigating the transitional period as we move from classical to quantum-resistant cryptographic systems. These solutions combine the best of both worlds by layering classical cryptographic algorithms, which are well-understood and efficient, with quantum-resistant algorithms that offer a safeguard against the emerging threats posed by quantum computing. This multi-pronged approach not only ensures immediate security against existing threats but also future-proofs Web3 systems against the quantum vulnerabilities that are on the horizon. In this subsection, we will explore the merits of hybrid cryptographic solutions, how they function, and their applicability in securing Web3 platforms.

A hybrid cryptographic system typically operates by using both classical and quantum-resistant algorithms in tandem. For example, a secure transaction on a blockchain could be encrypted using a classical algorithm like RSA or ECC and then further encrypted using a quantum-resistant algorithm like a lattice-based or hash-based method. This dual encryption offers immediate security through the classical algorithm, which is generally faster and requires less computational power, while also incorporating the robust, future-proof security offered by the quantum-resistant layer. Essentially, even if an attacker were to break the classical encryption using a quantum computer, the quantum-resistant layer would still stand as an unbreached barrier.

One of the most compelling advantages of this approach is its adaptability. As the field of quantum computing progresses, new vulnerabilities may be discovered in existing quantum-resistant algorithms, or entirely new quantum algorithms may be developed that pose unforeseen threats. A hybrid system offers the flexibility to easily replace or update the quantum-resistant layer as needed, without overhauling the entire cryptographic system. This is particularly beneficial in a decentralized Web3 environment, where making network-wide changes can be logistically challenging and time consuming.

Another advantage is the optimization of computational resources. Quantum-resistant algorithms often involve more complex mathematical operations, which can be computationally intensive. Using them exclusively could slow down transaction speeds and increase the computational load on nodes participating in the network. A hybrid approach allows the network to leverage the efficiency of classical algorithms while still maintaining a secure stance against quantum threats.

Implementing a hybrid cryptographic solution in a Web3 context would involve several key steps. First, a suitable quantum-resistant algorithm—or multiple algorithms for added security—would need to be identified and tested rigorously for potential vulnerabilities. Following this, the chosen quantum-resistant method would be integrated into the existing cryptographic framework. Given the decentralized nature of most Web3 platforms, this would likely require a period of community consultation and possibly a vote among network participants to approve the upgrade. Once consensus is reached, the new hybrid cryptographic system could be rolled out as part of a scheduled network update.

9.3.4 Case Studies

This section delves into several pioneering case studies that highlight the innovative approaches and technologies being adopted by various entities to address Quantum security in the Web3 ecosystem. From leveraging quantum-resistant cryptographic primitives to implementing advanced ledger databases, these case studies provide valuable insights into the efforts being made to future-proof Web3 technologies against the impending quantum computing era.

1. The Nervos Network is proactively preparing for the challenges posed by quantum computing by incorporating quantum-resistant cryptographic primitives in its Layer 1, the Common Knowledge Base (CKB) which is designed to be flexible and adaptable, allowing for the incorporation of quantum-resistant cryptographic primitives to secure the network against potential quantum computing threats. It is capable of upgrading its basic cryptographic primitives with new quantum-resistant ones without undergoing a hard fork, which is a contentious

process that could take many months or years and wreak havoc on the network (Nervos, 2023).

2. The Quantum-Resistant Ledger (QRL): Focuses on quantum-resistant cryptography using lattice-based cryptography and threshold signatures (Maha, 2023). QRL distinguishes itself in the blockchain and digital asset landscape through several unique features and offerings that address the growing concerns around quantum computing and its potential impact on cryptographic security. QRL employs the Extended Merkle Signature Scheme (XMSS), a digital signature scheme recognized for its post-quantum security. QRL offers a suite of products designed to cater to a wide range of users, including developers, everyday users, and data miners. For developers, QRL offers a rich and open API, along with an open development infrastructure. These features are crucial as they facilitate the creation of quantum-safe blockchain applications.

3. The Quantum Ledger Database (QLDB) introduced by Amazon on September 10th, 2019, was a ledger database that incorporates elements of blockchain technology (Grootenboer, 2019). QLDB is distinctive for its fully managed ledger capability, which can host multiple tables and maintains an immutable transaction journal that is cryptographically verifiable. The system operates under a centralized authority, which differentiates it from traditional blockchain models that are usually decentralized. The design of QLDB allows for the organization of data in multiple tables within a single ledger, facilitating the management of complex data structures. This feature can be advantageous for businesses and organizations that need to maintain detailed and interconnected data records. A key aspect of QLDB is its immutable transaction journal. This journal records all changes in a sequential and permanent manner, ensuring that each entry, once recorded, cannot be altered. This immutability is critical for scenarios where maintaining an unalterable historical record is important for verification and audit purposes.

   Cryptographic verifiability is another feature of QLDB. Each transaction and modification in the ledger is secured cryptographically, which adds a layer of security to the system and ensures that the data is protected against unauthorized changes. This feature supports the integrity and authenticity of the data stored in the ledger. QLDB's centralized management structure sets it apart from typical decentralized blockchain systems. This centralization can lead to more streamlined operations and management, addressing some of the challenges faced by decentralized systems, such as slower transaction speeds and the complexities of achieving consensus. Centralization also enables the overseeing authority to enforce standards and regulations more effectively although centralization may incur security and trust costs that decentralization aims to overcome.

4. JPMorgan Chase, in collaboration with Toshiba and Ciena, has pioneered the implementation of a QKD network in a metropolitan setting (Pistoia, 2023). This breakthrough, facilitated by JPMorgan Chase's Future Lab for Applied Research and Engineering (FLARE) and Global Network Infrastructure teams, showcases a QKD network resistant to quantum computing attacks and supporting data rates of 800 Gbps for critical applications under real-world conditions.

This achievement includes the successful deployment of QKD to secure a bank-led, production-grade, peer-to-peer blockchain network. The implementation marks the first instance of QKD being used to secure a vital blockchain application in the industry.

Key accomplishments in JPMorgan Chase's fiber optic production simulation lab, in collaboration with Toshiba and Ciena, include:

- The integration of a QKD channel with ultra-high bandwidth 800 Gbps optical channels on the same fiber, providing encryption keys for the data stream.
- Successful coexistence of the quantum channel with two 800 Gbps and eight 100 Gbps channels over a 70-km fiber, generating keys at a rate to support up to 258 AES-256 encrypted channels with a refresh rate of one key per second.
- Effective operation of QKD alongside ten high-bandwidth channels for distances up to 100 km.

The project utilized Toshiba's Multiplexed QKD System, produced in Cambridge, UK, and Ciena's Waveserver 5 platform, which features 800 Gbps optical-layer encryption and open APIs over Ciena's 6500 photonic solution.

The case studies presented illustrate the diverse and proactive strategies being employed by various Web3 entities to combat the security challenges posed by quantum computing. Nervos Network's integration of quantum-resistant cryptographic primitives into its Layer 1 platform, the Quantum-Resistant Ledger's adoption of lattice-based cryptography and threshold signatures, Amazon's Quantum Ledger Database with its unique centralized ledger system, and JPMorgan Chase's pioneering use of QKD in a metropolitan blockchain network exemplify the cutting-edge solutions being developed. These initiatives not only highlight the growing awareness of quantum threats but also showcase the commitment to innovation and adaptation within the Web3 community. By embracing such forward-thinking approaches, these projects set a benchmark for developing robust, quantum-resistant infrastructures, ensuring the longevity and security of Web3 technologies in a future dominated by quantum computing.

9.4 Conclusion

This chapter provides a comprehensive examination of the intersection of quantum computing and Web3 technologies. It highlights the profound implications that the advent of quantum computers poses to the security foundations of blockchain networks and the broader Web3 ecosystem. While quantum computing promises unmatched processing power, algorithms like Shor's threaten the viability of widely used cryptographic techniques. From the potential compromise of digital signatures and consensus protocols to the disruption of vital components like decentralized finance and non-fungible tokens, the risks present significant technical and economic challenges.

However, all is not lost, as active research and innovation focused on "quantum-resistant" algorithms provide hope for a smooth transition into the quantum era. By shifting to cryptographic techniques rooted in lattice-, hash-, and code-based problems, blockchains and Web3 systems can re-establish resilience against quantum attacks. Mirroring the decentralized ethos, a multi-faceted defense strategy including QKD, hybrid cryptography, and best security practices will strengthen the backbone of open, transparent and tamper-proof networks.

Progress lies in recognizing the gravity of the impending quantum revolution, actively collaborating across stakeholders, and proactively developing sophisticated cryptographic defenses to safeguard the promise of Web3. Incremental steps like algorithmic improvements, pilot testing in simulated environments and draft standards being pioneered today, must gather steam to ensure that quantum computing remains an avenue for advancing human potential rather than a harbinger of disruption. By reflecting on the opportunities and challenges outlined in this chapter, one hopes all actors can collectively steward blockchain networks into an era dominated by quantum—where decentralization and security continue to coexist in harmony.

References

Ahmad, K. (2022, December 8). *What Is Lattice-Based Cryptography and Why Is It Important?* MakeUseOf. Retrieved from https://www.makeuseof.com/what-is-lattice-based-cryptography/

Amazon Quantum Solutions Lab. (2023, November 13). *A detailed, end-to-end assessment of a quantum algorithm for portfolio optimization, released by Goldman Sachs and AWS | Amazon Web Services*. AWS. Retrieved from https://aws.amazon.com/blogs/quantum-computing/a--detailed-end-to-end-assessment-of-a-quantum-algorithm-for-portfolio-optimization-released-by-goldman-sachs-and-aws/

Baker, B. (2023a, August 31). *NIST Releases Draft Standards for Quantum-Resistant Algorithms—NIST Releases Draft Standards for Quantum-Resistant Algorithms*. AI Business. Retrieved from https://aibusiness.com/verticals/nist-releases-draft-standards-for-quantum-resistant-algorithms#close-modal

Baker, B. (2023b, September 5). *IonQ Targets Quantum Advantage via Generative AI*. Enter Quantum. Retrieved from https://www.quantumbusinessnews.com/research/ionq-targets-quantum-advantage-via-generative-ai

Brothwell, R. (2023, June 9). *5 exciting use cases for blockchain in government*. BSV Blockchain. Retrieved from https://www.bsvblockchain.org/news/5-exciting-use-cases-for-blockchain-in-government

Cook, J. (2019, March 23). *Code-based cryptography | McEliece PQC etc*. Applied Mathematics Consulting. Retrieved from https://www.johndcook.com/blog/2019/03/23/code-based-cryptography/

De Miguel, S. E. (2023, November 2). *Blockchain In The Legal Industry: Use Cases and New Legal Jobs*. Bigle Legal. Retrieved from https://blog.biglelegal.com/en/blockchain-in-legal-industry-use-cases-blockchain-jobs

de Quehen, V. (2020, February 24). *Math Paths to Quantum-safe Security: Hash-based Cryptography*. ISARA Corporation. Retrieved from https://www.isara.com/blog-posts/hash-based-cryptography.html.

GQI. (2023, March 25). *Tools of Quantum Computing—A List By Quantum Computing Report*. Quantum Computing Report. Retrieved from https://quantumcomputingreport.com/tools/

Grammel, G., Singh, S., Mitchell, C., & Joshi, A. (2021, June 10). *Quantum Key Distribution (QKD): How Does It Actually Work? Juniper Blogs.* Retrieved from https://blogs.juniper.net/en-us/security/quantum-key-distribution-qkd-how-does-it-actually-work

Greene, T. (2023, June 1). *Scientists propose quantum proof-of-work consensus for blockchain.* Cointelegraph. Retrieved from https://cointelegraph.com/news/scientists-propose-quantum-proof-of-work-consensus-for-blockchain

Grootenboer, E. (2019, September 17). *Amazon Announces General Availability of Quantum Ledger Database.* InfoQ. Retrieved from https://www.infoq.com/news/2019/09/amazon-quantum-ledger-database/

IBM. (2023, December 4). *IBM Debuts Next-Generation Quantum Processor & IBM Quantum System Two, Extends Roadmap to Advance Era of Quantum Utility.* IBM Newsroom. Retrieved from https://newsroom.ibm.com/2023-12-04-IBM-Debuts-Next-Generation-Quantum-Processor-IBM-Quantum-System-Two,-Extends-Roadmap-to-Advance-Era-of-Quantum-Utility

Lague, D. (2023, December 14). *U.S. and China race to shield secrets from quantum computers.* Reuters. Retrieved from https://www.reuters.com/investigates/special-report/us-china-tech-quantum/

Maha, M. (2023, January 12). *Quantum Resistant Ledger: A quantum-proof blockchain frontier.* ManageEngine. Retrieved from https://www.manageengine.com/active-directory-360/manage-and-protect-identities/identitude/blogs/quantum-resistance-ledger-QRL-and-cryptography-role-in-blockchain.html

Mastriani, M. (2021). *Quantum Fourier transform is the building block for creating entanglement.* NCBI. Retrieved from https://www.ncbi.nlm.nih.gov/pmc/articles/PMC8593191/

McKinsey. (2021, June 18). *Quantum computing in drug development.* McKinsey. Retrieved from https://www.mckinsey.com/industries/life-sciences/our-insights/pharmas-digital-rx-quantum-computing-in-drug-research-and-development

Microsoft. (2023, September 18). *Searching with Grover's Algorithm–Code Samples.* Microsoft Learn. Retrieved from https://learn.microsoft.com/en-us/samples/microsoft/quantum/searching-with-grovers-algorithm/

Nervos. (2023, June 5). *Quantum Resistance in Blockchains: Preparing for a Post-Quantum Computing World.* Nervos Network Retrieved from https://www.nervos.org/knowledge-base/quantum_resistance

NIST. (2023, August 24). *NIST to Standardize Encryption Algorithms That Can Resist Attack by Quantum Computers | NIST.* National Institute of Standards and Technology. Retrieved from https://www.nist.gov/news-events/news/2023/08/nist-standardize-encryption-algorithms-can-resist-attack-quantum-computers

Pavlidis, A., & Gizopoulos, D. (2022, July 19). *Quantum Cryptography–Shor's Algorithm Explained.* Classiq. Retrieved from https://www.classiq.io/insights/shors-algorithm-explained

Pistoia, M. (2023). *JPMorgan Chase, Toshiba and Ciena Build the First Quantum Key Distribution Network Used to Secure Mission-Critical Blockchain Application.* J.P. Morgan. Retrieved from https://www.jpmorgan.com/technology/technology-blog/jpmc-toshiba-ciena-build-first-quantum-key-distribution-network-critical-blockchain-application

Quantiki. (2015, October 26). *Shor's factoring algorithm.* Quantiki. Retrieved from https://www.quantiki.org/wiki/shors-factoring-algorithm

QuTech. (2022). *Superposition and entanglement.* Quantum Inspire. Retrieved from https://www.quantum-inspire.com/kbase/superposition-and-entanglement/

Rogucki, M. (2023, November 3). *Who is trying to build a quantum computer?* TS2 Space. Retrieved from https://ts2.space/en/who-is-trying-to-build-a-quantum-computer/#gsc.tab=0

Rouse, M. (2019, September 25). *What is Quantum Coherence?–Definition from Techopedia.* Techopedia. Retrieved from https://www.techopedia.com/definition/34025/quantum-coherence

SSL2BUY. (2022). *RSA vs ECC–Which is Better Algorithm for Security?* SSL2BUY. Retrieved from https://www.ssl2buy.com/wiki/rsa-vs-ecc-which-is-better-algorithm-for-security

Swayne, M. (2021, April 20). *DP World, D-Wave Using Quantum Computers to Optimize Supply Chain*. The Quantum Insider. Retrieved from https://thequantuminsider.com/2021/04/20/dp-world-d-wave-partner-to-explore-using-quantum-computers-to-optimize-supply-chain/

Swayne, M. (2023, March 24). *What Are The Remaining Challenges of Quantum Computing? The Quantum Insider*. Retrieved from https://thequantuminsider.com/2023/03/24/quantum-computing-challenges/

Thummalapenta, R. (2023, October 22). *Is Quantum Computing the End of Blockchain?|by Riti Thummalapenta|Oct, 2023*. Medium. Retrieved from https://medium.com/@riti.thummalapenta/is-quantum-computing-the-end-of-blockchain-f3734cf84331

Vijayakumar, P. (2022). *Blockchain technology in healthcare: A systematic review*. NCBI. Retrieved from https://www.ncbi.nlm.nih.gov/pmc/articles/PMC9000089/

Jerry Huang has worked as a technical and security staff at several prominent technology companies, gaining experience in areas like security, AI/ML, and large-scale infrastructure. At Metabase, an open-source business intelligence platform, he contributed features such as private key management and authentication solutions. As a Software Engineer at Glean, a Generative AI search startup, Jerry was one of three engineers responsible for large-scale GCP infrastructure powering text summarization, autocomplete, and search for over 100,000 enterprise users. Previously at TikTok, Jerry worked to design and build custom RPCs to model access control policies. And at Roblox, he was a Machine Learning/Software Engineering Intern focused on real-time text generation models. He gathered and cleaned a large multilingual corpus that significantly boosted model robustness. Jerry has also conducted extensive security and biometrics research as a Research Assistant at Georgia Tech's Institute for Information Security & Privacy. This resulted in a thesis on privacy-preserving biometric authentication. His academic background includes a BS/MS in Computer Science from Georgia Tech and he is currently pursuing an MS in Applied Mathematics at the University of Chicago. phone: 571–268-6923;

Ken Huang is the author and chief editor of eight books on Generative Artificial Intelligence and Web3, published respectively by international publishers including Springer, Cambridge University Press, John Wiley, and China Machine Press. He currently serves as the CEO of the AI and Web3 consulting and education company DistributedApps.AI, based in the United States. Additionally, he holds multiple roles including the expert member of the Blockchain Committee of the Chinese Institute of Electronics, the Co-Chair of AI Organization Responsibility Working Group at Cloud Security Alliance, and Chair of the Blockchain Security Working Group at the Cloud Security Alliance, GCR. He is also a core contributor to the Generative AI Working Group at the NIST and a core author of the OWASP Top 10 for LLM Applications.

Ken Huang has been invited to provide Speaking or Consulting services at institutions including the University of California, Berkeley, Stanford University, Peking University, Tsinghua University, Shanghai Jiao Tong University, China Pacific Insurance, and the World Bank in the past.

– Moreover, he has given keynote speeches at international conferences, such as:
– The Davos World Economic Forum 2020 Blockchain Conference
– Consensus 2018 in New York
– The American ACM AI & Blockchain Decentralized Annual Conference 2019
– IEEE Technology and Engineering Management Society Annual Meeting 2019
– Silicon Valley World Digital Currency Forum
– Sino-US Blockchain Summit in Silicon Valley

He has also been awarded the "Blockchain 60" Figure Award by the National University Artificial Intelligence and Big Data Innovation Alliance Blockchain Special Committee in China in 2021.

Chapter 10
Privacy-Preserving Computation and Web3

Jerry Huang (iD), **Ken Huang** (iD), **and Mudi Xu**

Abstract This chapter examines techniques for Privacy-Preserving computation that can be applied in blockchain and Web3 systems. It analyzes privacy issues at the blockchain network layer and explores privacy enhancing protocols. Core techniques discussed include homomorphic encryption, secure multiparty computation, zero-knowledge proofs, and differential privacy. Integration strategies are provided such as performance optimization, modular development, open standards, and enhancing user experience. The outlook highlights future threats like quantum computing and scale, matched by ongoing innovation across cryptographic protocols, Generative AI integration, and community collaboration.

As Web3 and blockchain technology continues its meteoric rise, an intricate tension has emerged around data—its immense power to transform industries through insights, and its innate pseudonymity with privacy implications. With each digital interaction generating trails of personal data, from preferences to transactions, upholding privacy is no longer just an option but an imperative. However, the decentralized ethos of Web3, while revolutionizing access and control, has inadvertently introduced complex privacy challenges. These range from transparency of transactions on public ledgers to potential vulnerabilities in smart contracts and decentralized applications.

The good news is that Privacy-Preserving Computation, with its array of cryptographic armor and mathematical magic, promises a way forward. Its techniques enable deriving insights from collective data without compromising individual

J. Huang
The University of Chicago, Chicago, IL, USA
e-mail: jerryh@uchicago.edu

K. Huang (✉)
DistributedApps LLC, Fairfax, Virginia, USA
e-mail: ken@distributedapps.ai

M. Xu
Cloud Security Alliance, Shanghai, People's Republic of China
e-mail: info@c-csa.cn

privacy. Computational methods can execute code on encrypted data, prove statements without revealing underlying information, and even introduce calibrated noise to mask individual data points. The key is applying these techniques judiciously within Web3's architecture to balance privacy with functionality.

This chapter offers a practical guide into Privacy-Preserving Computation techniques and their strategic integration in Web3 systems. Core mechanisms analyzed include Homomorphic Encryption for computing on encrypted data, Secure Multiparty Computation for collaborative analytics, Zero-Knowledge Proofs for validation without exposure, and Differential Privacy for collecting insights not individuals. Integration strategies discussed encompass everything from performance optimizations and design modularization to embracing standards and enhancing user experience.

10.1   Privacy in Blockchain Network Layer

The privacy of the blockchain network layer is critical for maintaining the security and confidentiality of the system. This section will examine some of the challenges associated with privacy using Bitcoin and Ethereum network layers as examples.

10.1.1   Analysis of Privacy at the Bitcoin Network Layer

In the world of blockchain, the security of the system is heavily impacted by the P2P network used to transport transactions and blocks. As such, it is important to analyze the P2P network topologies of popular blockchains like Bitcoin and Ethereum. Message encryption is also needed for confidentiality between peer-to-peer nodes in the blockchain network. Different blockchain networks leverage different peer-to-peer networks, with Bitcoin adopting an unstructured P2P network and Ethereum relying on the Kademlia DHT to manage its P2P network (See BoxKademlia DHT).

Kademlia DHT
Kademlia Distributed Hash Table (DHT) is a pivotal system in peer-to-peer (P2P) networks, known for its efficient data storage and retrieval (Stanford, 2021). Unique identifiers assigned to each node create a network where distances are measured using an XOR metric, facilitating a straightforward yet effective data organization. This method allows data, stored as key-value pairs, to be strategically placed on nodes closest to the data's key, ensuring rapid and efficient data access. Nodes maintain routing tables organized into "k-buckets," enhancing communication efficiency by providing detailed knowledge of closer nodes.

Kademlia's decentralized design offers remarkable fault tolerance. Regular updates to k-buckets allow nodes to adapt to network changes, maintaining robustness against failures and dynamic network conditions. This decentralization eliminates single points of failure, making Kademlia networks resilient.

Widely adopted in blockchain and file-sharing applications, Kademlia excels in these environments due to its scalability and efficiency. In blockchain, it aids in managing transaction histories and network states, while in file-sharing, it efficiently locates peers holding specific file fragments. The combination of efficiency, scalability, and decentralized architecture makes Kademlia a fundamental component in modern, large-scale P2P networks.

One major issue with blockchain is users' privacy concerns related to transaction graphs and IP address papping via Bayesian analysis (Olier, 2018). Forensics firms such as Chainalysis use probability heuristics models to trace the path of inputs and outputs of Bitcoin's UTXO transaction model. The common input ownership heuristic is used by these firms to extract data from large transaction clusters, but this is mainly confined to blockchain layer analysis. However, this assumption is not always true, as multiple entities can cooperate to create and sign a single transaction that spends inputs owned by multiple people, known as CoinJoin (Hoffman, 2023). Tools like Wasabi Wallet can be used to mitigate tracing, but network layer traffic analysis can still reveal the IP address of a transaction's origin (Wilmoth, 2021).

To address these privacy threats, Bitcoin developers and contributors have proposed Dandelion++, a network layer privacy enhancement that makes the broadcast routing path more convoluted to trace (Curran, 2018). Running Tor and a VPN can also reduce the ability of third-party observers to extract definitive information on transaction broadcasts, but sophisticated methods targeting MAC addresses for devices and unique browser fingerprints can present additional problems to users of overlay networks like Tor.

Anonymity and privacy are constantly evolving, and Bitcoin's community is continually addressing these issues, although they are often overlooked by mainstream users. Therefore, it is essential to continue analyzing and improving the P2P network topologies and message encryption of blockchain to ensure its security and protect users' privacy.

10.1.1.1 Dust Attack and Privacy Issues

A dusting attack is a tactic used by attackers to identify the individuals or groups behind bitcoin or crypto wallets by sending tiny amounts of cryptocurrency to them. The attack does not attempt to steal funds but rather focuses on breaking users' privacy. The name "dust" comes from the small amounts of crypto sent, like dust scattered across blockchain networks.

Malicious actors take advantage of the fact that most users do not notice small balance changes and monitor outflows of trace amounts to attempt to discover the user's identity and potentially blackmail them. However, the true danger of a dusting attack lies in its ability to monitor wallet activity and facilitate phishing attacks.

In 2018 and 2019, notable dusting attacks occurred on the Bitcoin and Litecoin networks (Kirova, 2023), respectively, with the attackers using the dust transactions to advertise their platforms. While no major consequences have been reported, it is difficult to track the aftermath of such attacks, and users may have fallen victim to them without realizing it. To avoid the negative consequences of dusting attacks, there are several methods that can be employed to prevent or mitigate their effects. It is important not to dismiss the potential harm of these attacks and to take steps to protect oneself.

Some platforms let you flag unspent transaction outputs (UTXO), and many exchanges offer the option to convert dust. You can also use a hierarchical deterministic wallet or an exchange that offers the conversion option to protect your privacy. If you're not sure what to do, using an exchange that offers the conversion option is the easiest choice. By doing these things, you can protect yourself from potential identity exposure through social engineering.

Dusting attacks are a unique form of attack on crypto users that aim to breach the privacy of recipients. While dusting attacks are not as serious as other attacks, it's still important to be careful. If you notice a small increase in your holdings, don't spend it. Instead, use the "convert dust" option if available on your platform.

10.1.1.2   Bitcoin Message Encryption Improvement Proposal

The Bitcoin network is vulnerable to privacy attacks due to the public nature of transaction data and the lack of encryption on P2P connections. To address these issues, the Bitcoin developer community has proposed the Bitcoin Improvement Proposal (BIP) 324, which aims to provide opportunistic transport encryption to conceal the transferred data and make the bytestream indistinguishable from random bytes to a passive eavesdropper. This makes privacy attacks more difficult and costly and makes them easier to detect (Sanak, 2019).

BIP 324 adds encryption for messages sent between Bitcoin peers using the streamcipherChaCha20(Nagaraj,2023)withaPoly1305MessageAuthenticationCode (Bernstein, 2021).

The proposed BIP 324 would enhance the security of Bitcoin transactions by adding a layer of encryption similar to HTTPS. The goal of BIP 324 is to keep private metadata from being revealed with regular Bitcoin transactions, providing improved privacy for users. Although there is no set timeline for the adoption of BIP 324, the support from the Bitcoin developer community demonstrates a focus on improving confidentiality in the network. Node operators can choose to run a BIP 324 node by following the instructions at Github (https://github.com/bitcoin/bitcoin/pull/24545 issue 1,167,366,626).

10.2 Ethereum Message Encryption at Network Layer

The Waku protocol by the open-source community aims to transition from the older EIP627 (Whisper protocol) to a new system focused on enhancing peer-to-peer messaging with privacy and security (Waku, 2023). Waku is designed to reduce reliance on centralized intermediaries in messaging, emphasizing features like sender anonymity and metadata protection. These features are integral in decreasing the potential linkage of messages to identifiable personal data.

Waku supports a range of communication types, from human-to-human to machine-to-machine, and is adaptable to various platforms including desktops, servers, and resource-restricted devices like mobile phones and browsers.

The evolution of Waku is marked by the release of Waku v2, which is a redevelopment based on the Noise Protocol framework (Vac, 2022). This framework provides a structured approach to secure communications, involving custom key exchange protocols. It facilitates secure exchanges using operations like Diffie-Hellman to derive shared secret keys, bolstering security through features like confidentiality and forward secrecy.

Waku v2 introduces enhancements such as pub/sub functionality over libp2p, contributing to scalability and efficiency in decentralized messaging. Additionally, it includes capabilities for retrieving historical messages for devices with intermittent online connectivity, adaptive nodes for various network conditions, bandwidth conservation for lighter nodes, and message encryption. Although it is not clear when Waku will be widely adopted by the Ethereum community, the overall message privacy at the network layer is needed to protect privacy of the Ethereum ecosystem.

10.3 Foundations of Privacy-Preserving Computation

Before delving into specific techniques, understanding the foundational principles of Privacy-Preserving computation sets the stage. This section offers insights into the importance of data privacy, challenges in ensuring it, and the overarching goals of Privacy-Preserving computational methods.

10.3.1 The Imperative of Data Privacy in Web3

In the age of information, data is undeniably the new currency. As digital interactions proliferate, vast swathes of data are generated, exchanged, and stored. Web3, with its decentralized ethos, stands at the forefront of this data revolution, promising greater autonomy, transparency, and control for users over their digital footprints.

Yet, while Web3 brings forth a myriad of opportunities, it also presents unprecedented challenges in ensuring data privacy, making the discourse around its preservation all the more critical.

Web3's architecture, characterized by its decentralized networks, distributed ledgers, and peer-to-peer interactions, is fundamentally different from the centralized models of Web2. Unlike traditional systems where a single entity holds and controls data, Web3 disperses data across multiple nodes. On the surface, this seems to offer enhanced privacy—after all, there's no single point of control or failure. However, the landscape is far more nuanced.

One of the paradoxes of blockchain technology, a cornerstone of Web3, is its transparency. While transactions are pseudonymous, they are also public. Every transaction, once recorded on a blockchain, is visible to anyone who wishes to inspect the ledger. Over time, with advanced data analytics and pattern recognition, it might become possible to de-anonymize users, linking their blockchain activities to real-world identities.

Moreover, smart contracts, while automating and enforcing agreements on the blockchain, might also inadvertently expose sensitive data. If not meticulously designed, a smart contract can become a vector for data leaks, especially when interacting with external data sources or off-chain systems.

Beyond the inherent architectural challenges, Web3 interfaces with a plethora of decentralized applications (dApps). These dApps, while operating on decentralized networks, might have varying degrees of privacy measures. An insecure dApp could jeopardize the data of its users, even if the underlying blockchain remains secure.

The implications of data breaches in the Web3 realm can be profound. Financial losses aside, users risk exposure of sensitive personal information, compromising their privacy and security. In decentralized finance (DeFi) platforms, for instance, transactional privacy is crucial. Revealing a user's financial activities could make them a target for malicious actors.

In essence, the decentralized promise of Web3, while empowering, also brings to the fore intricate privacy challenges. The very features that make it revolutionary—transparency, immutability, and decentralization—also demand innovative approaches to preserving data privacy. As Web3 continues to evolve, intertwining more deeply with our daily lives, the imperative for robust, holistic, and adaptive data privacy measures has never been greater. It's a call to action for all stakeholders—developers, users, and regulators—to collaboratively forge a path that upholds the sanctity of personal data in the decentralized digital age.

10.3.2 Balancing Utility with Confidentiality

The meteoric rise of Web3 has underscored a dilemma: the tug of war between data utility and confidentiality. As Web3 applications burgeon, offering a plethora of services ranging from decentralized finance (DeFi) to supply chain management, they inevitably leverage vast quantities of data. This data, while pivotal for functionality, also presents a potential goldmine for malicious actors, thereby highlighting the urgent need to strike a delicate balance between its utility and confidentiality.

At the heart of this balance is the recognition that data drives the modern digital economy. Web3 applications, by their very nature, leverage data to provide bespoke services, ensure seamless user experiences, and generate insights. For instance, a DeFi platform might analyze transactional data to offer personalized financial products, while a decentralized marketplace might use browsing patterns to curate product recommendations. In these contexts, data isn't just an ancillary component; it's the lifeblood that fuels functionality and innovation.

Yet, with this utility comes an inherent risk: the potential compromise of confidentiality. Every piece of data harnessed by a Web3 application is a potential point of vulnerability. And given that Web3 operates on decentralized, often transparent ledgers, the risks aren't just about unauthorized access but also about unintended exposure. A simple transaction on a blockchain might reveal more about a user than they'd like, from their spending habits to their associations.

Balancing this dichotomy necessitates a multipronged approach:

1. Data Minimization: One of the foundational principles of data privacy is only to collect what's absolutely necessary. Web3 applications can adopt a minimalist approach, ensuring that they only use data essential for their operations, thereby reducing potential exposure points.
2. Differential Privacy: Techniques like differential privacy can be employed to add "noise" to data, ensuring that while aggregate insights can be derived, individual data points remain obfuscated. We will discuss more about this in Sect. 10.4.4.
3. Zero-Knowledge Proofs: This cryptographic method allows one party to prove to another that a statement is true, without revealing any specific information about the statement itself. It's a promising tool for Web3 applications where transactional validity is crucial, but the specifics of the transaction need to remain confidential. Please refer to Sect. 10.4.3 for more details.
4. On-chain vs. Off-chain Data: Deciding what data resides on the blockchain and what stays off chain is a strategic decision. Sensitive data can be kept off chain, ensuring confidentiality, while reference hashes or pointers can be stored on chain to maintain integrity.
5. User Empowerment: At the end of the day, users should have control over their data. Providing them with tools to manage, revoke, or grant access ensures that they remain at the center of the data utility vs. confidentiality equation.

In essence, as Web3 continues its transformative journey, the balance between data utility and confidentiality will remain a pivotal concern. It's not just about technological solutions but also about ethical considerations, user empowerment, and proactive governance. The future of Web3, if it's to be both revolutionary and secure, hinges on this intricate balance, demanding continuous innovation, vigilance, and collaboration from the community.

10.3.3   Goals and Challenges
of Privacy-Preserving Computation

The ascendancy of Web3 has reshaped our digital landscape, introducing both unparalleled opportunities and intricate challenges. Among these, the pursuit of Privacy-Preserving computation stands out, reflecting a collective aspiration to harness the power of data while safeguarding individual privacy. To navigate this complex terrain, it's vital to delineate clear goals and recognize the challenges that lie in achieving them.

10.3.3.1   Goals of Privacy-Preserving Computation

1. Data Confidentiality: The foremost goal is to ensure that sensitive data remains confidential during computation. This means that even during processing, raw data remains obscured, preventing unauthorized access or inference.
2. Result Integrity: While data remains concealed, the accuracy of computational outcomes is paramount. Privacy-Preserving methods aim to produce results that are both correct and verifiable.
3. Computation Efficiency: The computation, even when preserving privacy, should remain efficient. This pertains to both time and computational resources, ensuring that privacy measures don't hinder performance unduly.
4. User Control: Empowering users to control their data, decide its use, and understand the computations performed on it is a foundational goal. Privacy isn't just about concealment but also about agency and user consent.
5. Interoperability: In the diverse world of Web3, different platforms, applications, and systems intermingle. Privacy-Preserving methods should, ideally, be interoperable across this landscape, ensuring broad applicability.

10.3.3.2   Challenges in Achieving these Goals

1. Computational Overhead: Privacy-Preserving techniques, especially cryptographic ones, can introduce computational overheads, making some operations slower or more resource-intensive.
2. Complex Implementation: Some methods, like homomorphic encryption or zero-knowledge proofs, can be intricate, demanding expertise and careful implementation to ensure both privacy and functionality.
3. Data Utility Trade-off: There's often a trade-off between the level of privacy ensured and the utility of data. Too much obfuscation might render data less useful for certain computations.
4. Regulatory and Compliance Hurdles: As privacy regulations evolve globally, adhering to diverse and sometimes conflicting norms can be challenging for Web3 platforms and applications.

5. User Education and Awareness: While tools and techniques evolve, a significant challenge lies in educating users about their choices, implications, and the nuances of Privacy-Preserving computations.
6. Coordination Across Entities: Web3's decentralized nature implies multiple stakeholders—developers, node operators, users, and more. Achieving consensus or coordinating Privacy-Preserving efforts across such a diverse ecosystem can be challenging.

Therefore, the quest for Privacy-Preserving computation in the Web3 milieu is both a technical and sociological endeavor. It calls for cutting-edge solutions, informed users, and a collaborative spirit. While the goals are clear, the path is laden with challenges, necessitating continuous innovation, dialogue, and a commitment to a future where data empowerment doesn't come at the cost of privacy.

10.4   Techniques in Privacy-Preserving Computation

Diverse techniques have been developed to address the multifaceted challenges of data privacy. This section delves deep into prominent methods, explaining their mechanisms, benefits, and potential applications in Web3.

10.4.1   Homomorphic Encryption

The world of cryptography has witnessed a paradigm shift with the introduction of homomorphic encryption. In essence, it's a transformative encryption method that uniquely allows computations to be performed directly on encrypted data, producing results that, when decrypted, align perfectly with what would have been obtained had the computations been done on the original, unencrypted data (Internet Society, 2023). This groundbreaking technique has been hailed for its potential to redefine the boundaries of secure computation, especially in environments like Web3 where data privacy is paramount.

Figure 10.1 gives a high-level depiction of how homophobic encryption works.

We can understand how homomorphic encryption works from the following components and processes:

1. Data: This is the original data that needs to be processed securely.
2. Encrypt: The data is first encrypted. This means it is turned into a form that can't be understood by anyone who doesn't have the key to decrypt it.
3. Encrypted Data: This is the data after it has been encrypted. It is secure and cannot be read or understood by unauthorized parties.
4. Process in Encrypted Form: This step is what makes Homomorphic Encryption special. The encrypted data can be processed or computations can be performed

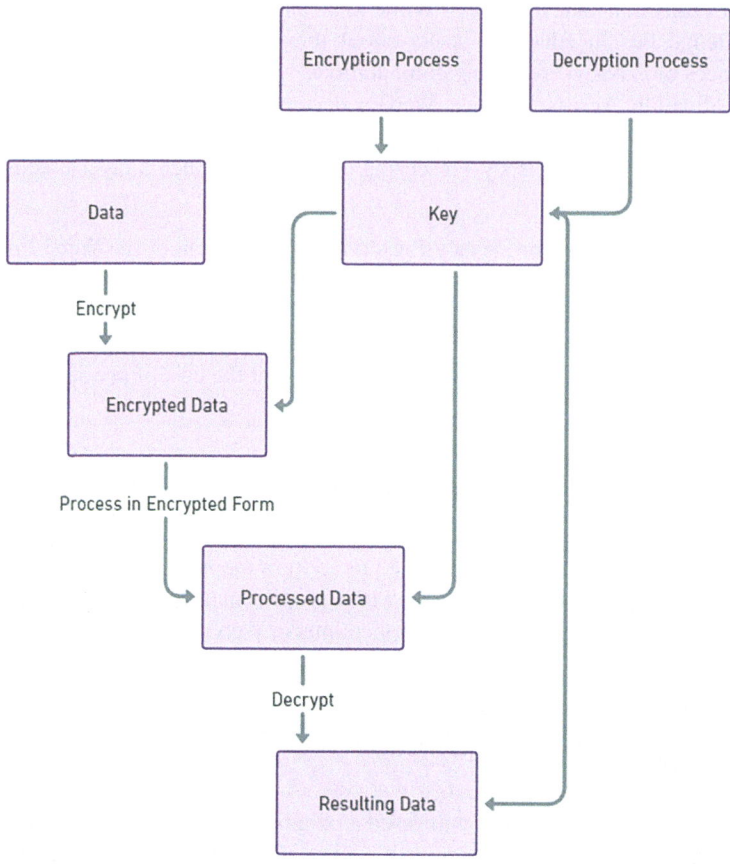

Fig. 10.1 How homomorphic encryption works

on it without ever decrypting it. This means the data stays secure even during processing.

5. Processed Data: This is the result of the computations performed on the encrypted data. It is still in an encrypted form and secure.

6. Decrypt (Only by Owner): After the processing is complete, the encrypted result can be decrypted, but only by the owner of the data or someone with the right decryption key. This ensures that only authorized individuals can access the final processed data.

7. Resulting Data: This is the data after it has been decrypted. It is now in a form that can be understood and used, but it has remained secure throughout the entire process.

8. Owner's Key: This key is used both for encrypting the original data and decrypting the processed data. It is crucial for the security of the data as only the owner, or someone with access to this key, can decrypt and access the final data.

The essence of Homomorphic Encryption is to allow data to be securely processed without ever exposing it in an unencrypted form, thereby preserving privacy and security.

Homomorphic encryption's foundational concept revolves around the ability to undertake meaningful operations on encrypted data, all without revealing the raw, sensitive data itself. Think of it as performing arithmetic on a locked treasure chest and, when you finally decide to open it, finding the exact treasures you would expect based on your calculations. The data remains continuously shielded, even during computational processes, drastically reducing exposure risks.

In the decentralized realm of Web3, homomorphic encryption's potential is particularly palpable. For instance, consider a scenario involving a decentralized database on a blockchain. Users might want to query this database without laying bare their queries or the underlying data. Homomorphic encryption can facilitate this, allowing computations that return encrypted results decipherable only by the original querier. Similarly, in the world of smart contracts this form of encryption can ensure these contracts evaluate encrypted conditions, never accessing the raw data, thereby ensuring a blend of operational efficiency and user privacy.

The benefits of homomorphic encryption are manifold. Beyond the evident enhancement of data privacy, it aids in ensuring compliance with the increasingly stringent global data protection regulations such as EU's GDPR (Wolford, 2022). Organizations can process data, confident in their adherence to privacy norms. Additionally, the flexibility offered by this encryption method is unparalleled. It can accommodate a plethora of operations, from basic arithmetic to more intricate functions, all the while keeping the data encrypted.

However, as with all innovations, it's not devoid of challenges. The computational overheads associated with homomorphic encryption are notably significant. Operations on encrypted data tend to be much slower than their plaintext counterparts, potentially impacting real-time applications. Furthermore, the intricacies of the underlying mathematics and its implementation necessitate specialized expertise. There's also the matter of noise accumulation, a phenomenon where, as operations are performed on encrypted data, a form of computational "noise" builds up, which, if unchecked, can skew results (Weinkauf, 2023). Parameter selection, vital for the efficiency and security of the encryption, is another complex aspect, demanding meticulous attention to detail (Chase et al., 2017).

Looking forward, the trajectory of homomorphic encryption seems promising. With ongoing research and open-source projects like Microsoft's SEAL (Microsoft, 2018) and IBM's HElib (IBM, 2022) striving to refine its efficiency and applicability, we're on the cusp of more widespread adoption. In the evolving landscape of Web3, tools like homomorphic encryption will undoubtedly take center stage, ensuring that as we venture into a decentralized digital future, we do not compromise on the sanctity of data privacy.

10.4.2 Secure Multiparty Computation (MPC)

In the age of data-driven decision making and interconnected digital systems, there's an ever increasing need to process data from multiple sources. However, the challenge lies in processing this collective data without exposing or compromising the confidentiality of individual datasets. Secure Multiparty Computation (MPC) emerges as a beacon in this landscape, promising collaborative data processing while upholding the sanctity of data privacy (Inpher, 2021).

Figure 10.2 represents the process of Secure Multiparty Computation (MPC), which is a way for multiple parties to compute a result together without revealing their private data to each other.

Here's how it works, in simple terms:

1. Private Data of Each Party: There are three parties involved, Party A, Party B, and Party C. Each party has its own private data that they want to keep secret.
2. Compute Partial Result: Each party independently computes a part of the result using their private data. They do this in a way that the partial result doesn't reveal their private data to others.
3. Combined Computation: All these partial results from Party A, Party B, and Party C are then combined together in a central computation. This central computation does not have access to the private data of the parties, just the partial results.

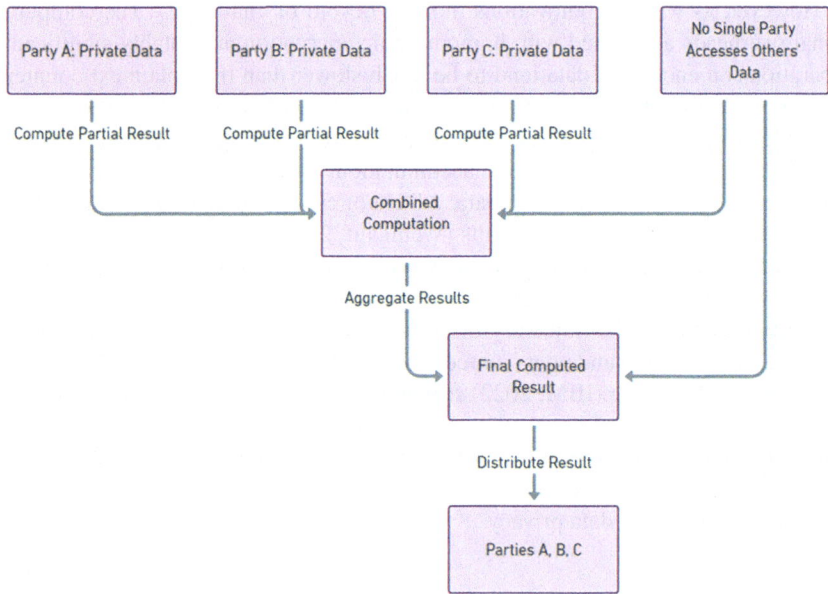

Fig. 10.2 Secure multiparty computation

4. Aggregate Results: The combined computation then aggregates or puts together these partial results to produce a final result.
5. Distribute Result: This final result is then shared with all parties. This final result doesn't reveal the private data of any individual party.
6. Privacy Preservation: Throughout this process, the key feature is that no single party can access the private data of the other parties. They all contribute to the final result without exposing their sensitive information.

This way, MPC allows for collaborative computation while maintaining the privacy and security of each party's data.

At its core, MPC is a cryptographic protocol that allows multiple parties to collaboratively compute a function over their inputs while keeping those inputs private. The beauty of this protocol is that, even though the computation might involve data from various sources, none of the participating entities get to see the other parties' actual data. They only witness the final computed result. It's akin to multiple individuals working on a joint project, each contributing their expertise, but without fully knowing what the others are specifically adding. In the end, they all witness the final product but remain unaware of the individual contributions.

In the context of Web3, the implications and applications of MPC are profound. The decentralized nature of Web3 platforms often requires various nodes or participants to reach consensus or make collective decisions. Traditional methods might necessitate sharing data across nodes, potentially exposing sensitive information. MPC, however, ensures that these computations or consensus decisions can be reached without any party revealing their data. Imagine a decentralized finance (DeFi) platform where multiple parties want to determine the average value of an asset they own. Using MPC, they can ascertain this average without ever revealing the exact value each party holds.

Moreover, MPC finds applications beyond just average computations. It can be used in auctions, voting systems, and even machine learning models where data from multiple sources is required to train a model, but the individual datasets must remain confidential. For instance, in a decentralized voting system on a blockchain, MPC can ensure that the final count is accurate without revealing individual votes.

MPC can also be used in Web3 wallets to enhance security by requiring multiple parties to accept, validate, and sign transactions. An MPC wallet leverages MPC technology to split the traditional private key into multiple shares, distributed among different parties, such as wallet users or trusted servers (Leal, 2023). This approach provides enhanced security, flexibility, and control over digital assets, making it a foundational piece of infrastructure for institutional custodians, investors, and traders (Alchemy, 2023). The use of MPC technology in wallets offers several advantages, including eliminating the need to trust a single party with the private key and providing enhanced security, risk mitigation, and more efficient asset management and transfer.

However, while MPC holds immense promise, it's essential to recognize that it's computationally intensive. Running these protocols requires robust computational power, and the complexity grows with the number of participants. Furthermore, ensuring robustness against malicious actors, who might try to skew the computation, is a challenge. Effective MPC protocols need to be resistant to such adversarial actions.

The intersection of MPC with Web3 technologies heralds a future where collaboration and privacy aren't mutually exclusive. As Web3 platforms grow in number and complexity, integrating protocols like MPC will be pivotal in ensuring that these decentralized systems can function effectively without compromising on the core tenet of individual data privacy. As we move toward a more interconnected digital world, tools like MPC will undoubtedly play a crucial role in shaping a future where collaboration is secure, and privacy is sacrosanct.

10.4.3 Zero-Knowledge Proofs

In the vast expanse of cryptographic techniques, zero-knowledge proofs (ZKPs) stand out as one of the most intriguing and transformative. At a high level, they offer an astonishing capability: allowing one party to prove to another that a given statement is true, without conveying any additional information apart from the veracity of the statement itself. In other words, they validate knowledge without revealing it.

Figure 10.3 gives a high-level process diagram of ZKP.

Zero-Knowledge Proofs are like a magic trick where someone (the Prover) convinces another person (the Verifier) that they know a secret, without actually revealing what the secret is. Here's how the process works:

1. Prover has Secret Information: This is someone who knows a secret and wants to prove that they know it without giving it away.
2. Zero-Knowledge Proof Process: The Prover goes through a special process that generates proof. This proof is a way of saying "I know the secret" without actually showing what the secret is.
3. Generates Proof Without Revealing Secret: The Prover sends this proof to the Verifier. Importantly, this proof doesn't contain any information about the secret itself.
4. Verifier Checks Proof: The Verifier receives the proof and checks it. They don't learn anything about the secret itself, but they can confirm whether the Prover really knows the secret.
5. Convinced or Not Convinced: After checking, the Verifier is either convinced that the Prover knows the secret (without knowing the secret themselves), or they are not convinced.
6. No Secret Information Leaked: Throughout this entire process, the actual secret is never revealed or leaked. The Verifier learns nothing about the secret, just that the Prover knows it.

Zero-Knowledge Proofs are a way of proving knowledge or possession of information without revealing the information itself, ensuring privacy and security.

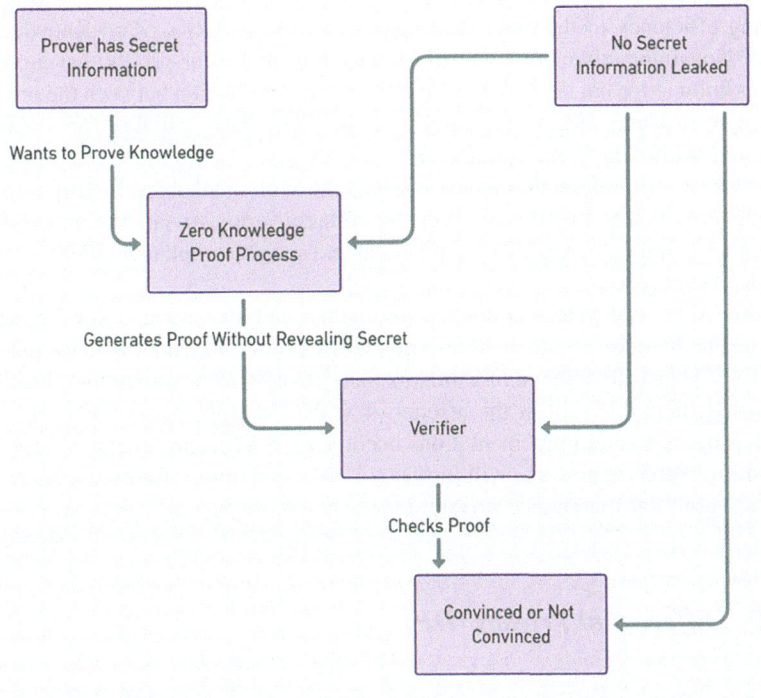

Fig. 10.3 Zero-knowledge proof process

To fathom the essence of zero-knowledge proofs, envision a scenario where Alice wants to prove to Bob that she knows a secret password, but she doesn't want to disclose the password itself. Through ZKPs, Alice can convince Bob of her knowledge without ever revealing the actual secret. This seemingly magical property has profound implications, especially in a domain like Web3, where data privacy and validation are paramount.

The power of ZKPs in Web3 becomes evident when we consider scenarios like transaction validation on blockchains. In a typical blockchain transaction, details like sender, receiver, and amount are recorded on a public ledger. While this transparency is a strength, it might not be suitable for all applications due to privacy concerns. Enter ZKPs, which can validate the correctness of a transaction without revealing the details of the transaction itself. This way, the blockchain can confirm that funds were indeed transferred without displaying the involved parties or the amount.

Another compelling application in the Web3 space is in identity verification. Digital identities are crucial in many Web3 applications, but users might not want to expose their entire identity or associated attributes. With ZKPs, users can prove specific attributes about themselves (e.g., being over a certain age) without revealing other details or the attribute's exact value.

Yet, the marvel of ZKPs doesn't stop at their ability to conceal information. They also bring efficiency to the table. Techniques like zk-SNARKs (Zero-Knowledge Succinct Non-Interactive Argument of Knowledge) allow for proofs that are not only zero knowledge but also succinct and require no interaction between the prover and verifier (Rasure, 2021). This makes them particularly suitable for scalable applications within the Web3 framework.

However, as with all technologies, ZKPs come with challenges. Crafting zero-knowledge proofs that are both succinct and efficient is non-trivial. The underlying mathematics is intricate, and creating practical and scalable implementations necessitates deep expertise.

As the Web3 ecosystem continues to mature and the demand for Privacy-Preserving techniques amplifies, zero-knowledge proofs will undoubtedly play a pivotal role. They offer a blend of validation and privacy that is rare to find, making them an indispensable tool in the arsenal of Web3 developers and architects. As research progresses and implementations become more efficient, the integration of ZKPs within Web3 frameworks will further solidify, anchoring a future where transactions are both transparent and private.

10.4.4 Differential Privacy (DP)

Differential Privacy (DP) provides a viable framework that ensures data privacy when extracting insights from large datasets (Devaux, 2022). Differential Privacy operates on a foundational premise: statistical queries to a database should not reveal whether a specific individual's information is included in it. To put it another way, the outcome of any analysis performed on a dataset should be nearly identical, irrespective of the participation of any single individual. This ensures that while aggregate information can be derived, the data about a particular individual remains indistinguishable.

As shown in Fig. 10.4, the magic of DP lies in the introduction of "noise" or randomness to the results of queries made on a dataset. This noise, while ensuring privacy, is calibrated in a way that the overall accuracy of the query's result remains largely intact. Essentially, DP strikes a delicate balance between data utility and data privacy. The degree of noise introduced is often governed by a parameter, typically referred to as epsilon (ε), which quantifies the privacy loss in the data. A lower value of ε indicates better privacy guarantees, but also implies that more noise has been added, potentially affecting the utility of the data.

The relevance of Differential Privacy in the Web3 paradigm is profound. As decentralized applications and platforms handle increasing volumes of user data, the need to extract insights without compromising on individual privacy becomes paramount. For instance, a decentralized marketplace might want to analyze purchasing patterns to improve its services, but without revealing data about individual purchases. Here, employing differential privacy can allow the marketplace to gather

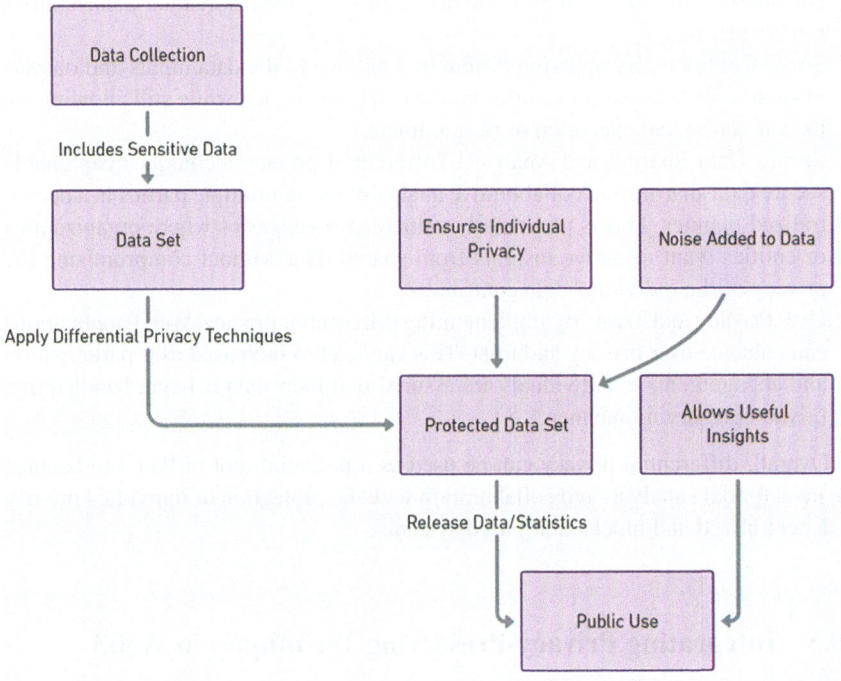

Fig. 10.4 Differential privacy

useful aggregate insights while ensuring that the specifics of any single user's trans-actions remain obscured.

Moreover, in a world increasingly wary of data breaches and misuse, DP offers a way for organizations to share data without the risks associated with exposing raw, individual level data. This is especially useful for decentralized platforms that might want to collaborate or share datasets for mutual benefits. By applying DP, they can share data-derived insights without ever risking the exposure of individual user data.

However, implementing Differential Privacy is not without challenges. Choosing the right value of ε, which offers a trade-off between privacy and accuracy, is a nuanced decision. Additionally, ensuring that the added noise doesn't significantly compromise the utility of the data requires careful calibration. Finally, while DP protects against certain types of data inference, it's not a silver bullet and needs to be combined with other Privacy-Preserving techniques described in this section for comprehensive protection.

Although not used in a real-world web3 project yet, DP can be applied to Web3 in several ways to ensure the confidentiality of sensitive data in decentralized and blockchain-based systems. Here are some key applications:

1. Decentralized Applications (dApps): Differential privacy can be integrated into the design of dApps to protect user data. This allows for the secure collection

and analysis of data from multiple users, ensuring that individual contributions remain private.

2. Smart Contracts: By applying differential privacy to the data inputs and outputs of smart contracts, sensitive information can be protected while still allowing for the validation and execution of the contracts.

3. Secure Data Sharing and Analysis: Differential privacy techniques can enable secure data sharing and collaborative analysis among multiple parties in a decentralized manner. This is particularly valuable for scenarios where organizations or entities want to derive insights from shared data without compromising the privacy of the individual data contributors.

4. User Privacy and Trust: By implementing differential privacy, Web3 applications can enhance user privacy and trust. This can lead to increased user participation and engagement, as individuals are assured that their data is being handled in a privacy-preserving manner.

Overall, differential privacy can be used as a potential tool in Web3 to balance the need for data analysis and collaboration with the protection of individual privacy in decentralized and blockchain-based systems.

10.5 Integrating Privacy-Preserving Techniques in Web3

With a plethora of Privacy-Preserving techniques available, their integration within Web3 applications requires strategic consideration. This section discusses best practices and the future outlook of Privacy-Preserving computation in Web3.

10.5.1 Selecting the Right Technique

Given the assortment of Privacy-Preserving techniques available, determining the right fit for a specific Web3 application can seem like navigating a labyrinth. However, making an informed decision is highly recommended, as the chosen technique can significantly influence the application's performance, user trust, and overall success. This section discusses some strategies in choosing the right technique.

10.5.1.1 Determine Selection Criteria

We recommend the following selection criteria:

Data Type Considerations in Privacy Techniques
Firstly, the nature of the data in Web3 applications can vary widely, from numerical and categorical to complex, decentralized identities and cryptographic proofs. Certain privacy techniques are inherently more effective for specific data types. For

example, zero-knowledge proofs are particularly adept at handling cryptographic data in a privacy-preserving manner.

Application Purpose and Privacy Needs
Secondly, the purpose of the application in the Web3 space can greatly influence the choice of privacy technique. Whether the application is focused on analytics, transactions, or interactive elements such as decentralized autonomous organizations (DAOs) and gaming will dictate the privacy needs. Analytical applications might prioritize data aggregation techniques that preserve privacy, while transactional applications may rely more on secure multiparty computation or homomorphic encryption.

User Expectations in Web3 Applications
User expectations in the Web3 realm also play a pivotal role. Given the ethos of decentralization and self-sovereignty in Web3, users often have high expectations for privacy and security. This is especially true for applications dealing with sensitive financial transactions or personal data on the blockchain. Therefore, selecting techniques that align with these heightened privacy expectations is essential.

Regulatory Compliance and Privacy Techniques
The regulatory landscape is another critical factor, with emerging laws and regulations around blockchain technology and cryptocurrencies. Compliance with global data protection laws such as GDPR in the EU, or specific blockchain and cryptocurrency regulations, can influence the choice of privacy techniques. Ensuring that the chosen methods are adaptable to comply with evolving regulations is vital.

Scalability in Web3 Privacy Techniques
Lastly, scalability is a significant consideration in the Web3 space. The chosen privacy-preserving methods must be scalable to support potentially large and growing numbers of users typical of successful blockchain and decentralized applications. Techniques that provide privacy while maintaining the high performance and scalability required by these applications are ideal.

Therefore, the selection of privacy-preserving techniques in Web3 requires careful consideration of the data type, application purpose, user expectations, regulatory requirements, and scalability needs. The decentralized and often public nature of blockchain-based systems, combined with the unique types of data and interactions in Web3, demands a nuanced approach to privacy that balances user expectations, regulatory compliance, and technical feasibility.

10.5.1.2 Weighing the Techniques

Having established the criteria, it's essential to weigh the benefits and limitations of the primary Privacy-Preserving techniques:

Homomorphic Encryption (HE): HE allows computations on encrypted data without requiring decryption. For applications that need to perform complex calculations on private data, like financial or health analytics, HE can be invaluable.

However, HE can be computationally intensive, potentially impacting application performance, especially if real-time results are expected.

Secure Multiparty Computation (MPC): When multiple entities need to collaborate without revealing individual data sets, MPC shines. It's particularly useful for consortium blockchains or collaborative Web3 platforms. Yet, it necessitates robust communication channels, which can be a bottleneck for some applications.

Zero-Knowledge Proofs (ZKPs): ZKPs are perfect for applications where validation is essential, but data revelation isn't. They're particularly relevant for transaction validation in Web3 platforms. While they provide robust privacy assurances, constructing efficient ZKP systems can be complex.

Differential Privacy (DP): For applications looking to share insights without revealing individual data points, like decentralized data marketplaces or analytics platforms, DP is a strong contender. However, calibrating the right amount of noise to maintain data utility can be challenging.

10.5.2   Overcoming Integration Challenges

Integrating privacy-preserving techniques into Web3 applications is a challenging yet meaningful endeavor. The obstacles encountered in this process, though significant, are not insurmountable. This section discusses some strategies of overcoming the challenges.

10.5.2.1   Performance Optimization Strategies

One effective strategy for performance optimization is the use of parallel processing. This approach involves breaking down tasks into smaller, more manageable subtasks that can be executed simultaneously. In the context of blockchain and Web3, this technique is particularly useful for computations related to privacy-preserving algorithms like zero-knowledge proofs. These proofs, often computationally intensive, can benefit significantly from parallel processing, where different components of the computation are handled concurrently, leading to faster overall execution times. This not only speeds up transactions but also enhances the user experience by reducing waiting times.

Efficient algorithms form another cornerstone of performance optimization in privacy-preserving Web3 applications. The choice of algorithms, especially in cryptographic operations, plays a significant role in determining the performance of the application. For instance, homomorphic encryption allows computations on encrypted data, but traditional forms of this encryption can be very resource-intensive. By opting for more efficient forms of homomorphic encryption, or by using advanced cryptographic techniques like elliptic curve cryptography, developers can reduce the computational overhead, thereby enhancing the application's performance without compromising security or privacy.

Another key strategy is the offloading of tasks to more powerful external systems. This is particularly relevant for operations that are too resource-intensive for the average user's device or for the inherent limitations of a blockchain network. Complex computations, such as those needed for large-scale data analysis or advanced privacy-preserving calculations, can be offloaded to cloud-based services or nodes equipped with higher processing power. This approach not only improves performance but also allows for scaling up the application as needed. It's crucial, however, to ensure that this offloading does not compromise the privacy and security of the data. This might involve using secure, encrypted channels for data transfer or ensuring that the external systems adhere to stringent privacy standards.

10.5.2.2 Modular Development and Testing

Modular development is effective in managing the complexity inherent in privacy-preserving Web3 applications. By structuring these applications into smaller, independent modules, each dedicated to a specific aspect of privacy, the development process becomes more organized and manageable. This modular approach offers several advantages in the context of complex systems like those found in Web3.

Firstly, it allows developers to isolate specific functionalities or privacy features, making it easier to identify and address potential issues. For instance, one module might handle encryption, another could manage identity verification, and a third might deal with data storage. This separation of concerns means that changes or updates in one area, such as improving the encryption algorithm, can be made without impacting other parts of the application. It also simplifies the process of integrating new privacy features or adapting to evolving privacy regulations, as changes can be confined to relevant modules without necessitating a complete overhaul of the entire application.

In addition to simplifying development, modular design facilitates more efficient and focused testing. Each module can be tested independently for its specific functionalities, ensuring that each aspect of privacy is thoroughly vetted. This is particularly important in the realm of Web3, where the security and privacy implications are significant. Rigorous testing of each module ensures that the privacy measures implemented not only function correctly but also align with the intended privacy goals.

Moreover, modular development helps in maintaining the utility and accessibility of data. Privacy-preserving measures, while crucial, should not render the data useless or inaccessible for legitimate purposes. Through modular design, developers can fine-tune the balance between privacy and utility. For example, a module responsible for data access control can be designed to ensure that data is accessible to authorized parties in a secure manner, thereby maintaining data utility while preserving privacy.

The testing and refinement phase in modular development is an essential step. This phase involves thorough testing of each privacy-preserving module to ensure it functions as intended and aligns with the overall system requirements. The refinement process includes iterating on the design based on test results, user feedback, and changing requirements, ensuring that the privacy measures are not only effective but also user-friendly and compliant with relevant privacy standards and regulations.

The combination of modular development and rigorous testing results in Web3 applications that are more robust, secure, and adaptable. It allows for a more agile response to the ever-evolving landscape of privacy concerns and technological advancements in the blockchain space. By adopting a modular approach and placing a strong emphasis on testing and refinement, developers can create privacy-preserving Web3 applications that not only protect user privacy but also provide a seamless and effective user experience.

10.5.2.3 Interoperability and Open Standards

In Web3 development, achieving interoperability while maintaining privacy can be a nuanced and intricate task. It requires a careful blend of adopting open standards and fostering collaboration within the community. The goal is to ensure that various Web3 platforms and tools can operate in unison, offering a seamless experience to users, without compromising on privacy.

Open standards are the cornerstone of interoperability in Web3. Utilizing established protocols, like ERC-20 for fungible tokens and ERC-721 for non-fungible tokens, ensures that digital assets and transactions are compatible across different blockchain platforms. This standardization is crucial for developers aiming to build privacy-preserving applications that can interact with a wide array of platforms without encountering compatibility issues. Moreover, the use of standardized smart contract interfaces across different blockchains facilitates the secure and predictable interaction of contracts, which is vital for maintaining privacy.

Active collaboration within the Web3 community is equally important. Developers must engage in cross-platform development teams, sharing insights and strategies to overcome the unique challenges posed by different blockchain ecosystems. This kind of collaborative effort not only enhances the collective understanding of privacy-preserving techniques but also leads to the development of more robust and interoperable solutions. Contributing to open-source projects is a practical way to achieve this, as it allows for the pooling of resources and knowledge, ensuring that privacy-preserving solutions are tested, refined, and validated by a broad community of developers.

Incorporating privacy-preserving measures in a way that they are compatible across various platforms requires a strategic approach. Technologies such as zero-knowledge proofs (ZKPs) are instrumental in this regard. ZKPs enable the verification of transactions or data without revealing the underlying information, which is a cornerstone for privacy in blockchain applications. Designing ZKPs to be platform-agnostic ensures that they can be integrated into diverse blockchain architectures, thus enhancing privacy without hindering interoperability.

Additionally, cross-chain technologies like Polkadot or Cosmos play a pivotal role in enhancing interoperability. These technologies allow different blockchains to communicate and transfer value securely, which is particularly significant for applications that operate across multiple chains. However, implementing these technologies demands careful consideration to ensure that privacy is not compromised during cross-chain interactions.

Blockchain oracles, which serve as bridges between blockchains and the external world, need to be utilized with a privacy-focused approach. While they are essential for bringing external data onto blockchain platforms, it's crucial to design these oracles in a manner that upholds the privacy of the data and the users involved.

10.5.2.4   User Experience and Trust

Focusing on user experience and trust in the context of Web3 and privacy-preserving applications requires a holistic approach, blending technical excellence with transparent and user-friendly communication. The objective is to create an environment where users feel confident about how their data is managed and protected, and where their interactions with the application are intuitive and satisfying.

The technical implementation of privacy-preserving features must be seamless and unobtrusive from the user's perspective. This means that while the underlying technology might be complex, involving sophisticated cryptographic techniques or blockchain protocols, the user interface needs to be straightforward and easy to navigate. The aim is to mask the complexity of the technology while providing users with a smooth and efficient experience. For instance, when integrating blockchain transactions or data encryption methods into an application, the user should be able to execute actions with simple clicks or commands, without needing to understand the underlying processes.

Clear communication with users about how their data is being protected is vital. This involves more than just providing a privacy policy or terms of service; it requires an effort to explain in clear, understandable language what data is being collected, how it is being used, and what measures are in place to protect it. For Web3 applications, where data might be stored on a blockchain or managed through smart contracts, it's important to inform users about the benefits and limitations of these technologies in terms of privacy. This can be achieved through in-app guides, FAQs, or even interactive tutorials that help users understand the privacy aspects of the application.

Integrating user feedback mechanisms is another key strategy to enhance trust. By providing users with a way to give feedback, report issues, or suggest improvements, developers can create a sense of community and shared ownership. This feedback loop is not only crucial for identifying and addressing user concerns but also for continuously improving the application. In the context of Web3, where the technology is still evolving, user feedback can provide invaluable insights into real-world usage and expectations.

Moreover, demonstrating responsiveness to user feedback builds trust. When users see that their input is valued and leads to tangible improvements or changes in the application, their confidence in the application increases. This is especially important in the area of privacy, where users are often concerned about how new technologies handle their personal information.

Incorporating these elements into Web3 applications results in a more user-centric approach, where technical robustness is matched with transparency and receptiveness to user needs. By focusing on creating a positive user experience and fostering trust through clear communication and responsiveness, developers can encourage wider adoption and more meaningful engagement with their applications. This approach not only benefits users but also contributes to the overall health and sustainability of the Web3 ecosystem.

10.5.3 Future Outlook: Evolving Threats and Solutions

The digital realm is constantly in flux, with each passing moment bringing forth novel technologies, paradigms, and associated challenges. As Web3 further embeds itself into the global digital fabric, the realm of Privacy-Preserving computation will undoubtedly find itself at the crossroads of rapid innovation and escalating threats.

First and foremost, as Web3 applications proliferate, the amount of data coursing through decentralized networks will see exponential growth. The sheer volume and variety of this data will present an enticing treasure trove for malicious actors. The granularity of data, spanning from personal preferences to financial transactions, will demand even more robust privacy measures. This increase in volume will also challenge the scalability of existing Privacy-Preserving techniques. Innovations will be required to ensure that privacy measures can handle vast datasets without compromising on performance or security.

Concurrently, quantum computing, a domain that is rapidly moving from theoretical to practical, poses a significant threat to many cryptographic techniques underpinning privacy solutions today. As quantum computers inch closer to achieving supremacy, the cryptographic bedrock of many Privacy-Preserving mechanisms might be at risk. The future will likely see an increased emphasis on developing quantum-resistant cryptographic methods to ensure that privacy solutions remain inviolable in the face of quantum threats. Please refer to Chap. 9 of this book about Web3 and Quantum Attacks.

However, it's not just external threats that the domain will grapple with. As Web3 applications become more sophisticated, the interplay between different decentralized applications (dApps) will grow. These interactions, while enabling richer functionalities, could inadvertently leak private information if not handled with care. Hence, future Privacy-Preserving solutions will need to cater to not just individual applications but also the ecosystem's collective interactivity.

On the brighter side, the evolution of threats is matched by relentless innovation in defense. Techniques that are nascent today, such as multiparty computation or advanced zero-knowledge proofs, will mature, offering more efficient and versatile solutions.

Moreover, the integration of Generative AI into the realm of privacy-preserving techniques in Web3 could mark a significant shift in how privacy is managed and maintained. Generative AI, renowned for its advanced reasoning and generation capabilities, offers a dynamic approach to privacy protection that extends far beyond traditional methods. Generative AI, at its core, is capable of analyzing vast datasets and identifying complex patterns, a feature that can be leveraged to proactively assess potential privacy risks. By continuously monitoring data transactions and interactions within a blockchain network, this AI can detect anomalies or unusual patterns that may signify a breach or a threat to privacy. This kind of real-time analysis is invaluable in the ever-evolving landscape of cybersecurity, where threats are becoming increasingly sophisticated and harder to detect with conventional methods. Furthermore, Generative AI can play a crucial role in optimizing the application of privacy-preserving techniques. Based on its analysis of ongoing threats and vulnerabilities, the AI can recommend or automatically implement the most effective privacy-preserving strategies for a given scenario. For instance, it could suggest the use of more robust encryption methods in response to heightened risk levels, or it might propose the deployment of zero-knowledge proofs in situations where data authenticity needs to be verified without revealing the underlying information. Another compelling aspect of integrating Generative AI is its ability to learn and adapt over time. As it processes more data and encounters a broader range of privacy challenges, its algorithms can evolve, becoming more adept at predicting and mitigating potential breaches. This continuous learning process ensures that privacy-preserving techniques are not static but are instead constantly improving, adapting to new threats as they arise. Finally, Generative AI could potentially generate and simulate various privacy breach scenarios to better prepare Web3 applications for real-world attacks. By creating realistic models of potential security threats, developers can test and strengthen their privacy-preserving measures proactively, ensuring that they are robust enough to withstand actual attacks.

Furthermore, the Web3 community's collaborative spirit will be its most significant asset. Open-source endeavors, transparent research, and shared learnings will accelerate the development of robust privacy solutions. As challenges arise, the community will rally, pooling its collective expertise to devise innovative countermeasures.

In conclusion, the future of Privacy-Preserving computation in Web3 is teeming with possibilities and challenges. While threats will evolve, so will the defense mechanisms, driven by a global community's collaborative spirit and commitment to safeguarding user privacy. The road ahead is undeniably complex, but with shared determination and continuous innovation, a future where Web3 applications uphold the highest standards of privacy is within reach.

10.6 Conclusion

As Web3 continues to transform industries with its decentralized architecture, upholding privacy stands paramount. Myriad techniques like homomorphic encryption, secure multiparty computation, zero-knowledge proofs, and differential privacy empower deriving insights from data without compromising individual privacy. Strategic integration of these techniques can optimize performance, ensure regulatory compliance, and enhance user trust in Web3 systems.

However, evolving threats like quantum computing and exponentially increasing data volumes will challenge even robust privacy protocols. Continued innovation in cryptographic techniques, paired with judicious application of Generative AI and relentless collaboration across the Web3 community, offer hopeful countermeasures.

The path ahead will demand persistent vigilance and shared responsibility among stakeholders. As data underpins functionality, and transparency enables trust, balancing these pillars without sacrificing privacy remains intricate yet imperative. Overall, with patient perseverance, ethical considerations, and collective ingenuity, the Web3 ecosystem can anchor a digital future where decentralization and self-sovereignty thrive in harmony.

References

Alchemy. (2023). *What is a multi-party computation (MPC) wallet?* Alchemy. Retrieved December 20, 2023, from https://www.alchemy.com/overviews/mpc-wallet

Bernstein, D. J. (2021). *Poly1305-AES: a state-of-the-art message-authentication code.* DJ Bernstein's. Retrieved December 17, 2023, from https://cr.yp.to/mac.html

Chase, M., Chen, H., Ding, J., & Goldwasser, S. (2017). *Security of homomorphic encryption.* Microsoft. Retrieved December 20, 2023, from https://www.microsoft.com/en-us/research/wp-content/uploads/2018/01/security_homomorphic_encryption_white_paper.pdf

Curran, B. (2018, October 4). *What is The Dandelion Protocol? Complete Beginner's Guide.* Blockonomi. Retrieved December 17, 2023, from https://blockonomi.com/dandelion-protocol/

Devaux, E. (2022, December 21). *What is Differential Privacy: definition, mechanisms, and examples.* Statice. Retrieved December 20, 2023, from https://www.statice.ai/post/what-is-differential-privacy-definition-mechanisms-examples

Hoffman, N. (2023, August 17). *Trezor Expands Privacy Features, Introduces Coinjoin For Trezor Model One.* Bitcoin Magazine. Retrieved December 17, 2023, from https://bitcoinmagazine.com/business/trezor-expands-privacy-features-introduces-coinjoin-for-trezor-model-one

IBM. (2022). *homenc/HElib: HElib is an open-source software library that implements homomorphic encryption. It supports the BGV scheme with bootstrapping and the Approximate Number CKKS scheme. HElib also includes optimizations for efficient homomorphic ...* GitHub. Retrieved December 20, 2023, from https://github.com/homenc/HElib

Inpher. (2021). *What is Secure Multiparty Computation? – SMPC/MPC Explained.* Inpher. Retrieved December 20, 2023, from https://inpher.io/technology/what-is-secure-multiparty-computation/

Internet Society. (2023, March 9). *Homomorphic Encryption: What Is It, and Why Does It Matter?* Internet Society. Retrieved December 20, 2023, from https://www.internetsociety.org/resources/doc/2023/homomorphic-encryption/

Kirova, D. (2023, June 13). *Crypto Dust Attacks Explained*. SimpleSwap. Retrieved December 17, 2023, from https://simpleswap.io/blog/crypto-dust-attacks-explained

Leal, J. (2023, April 7). *What is an MPC Wallet? The Complete Guide [2023]*. thirdweb blog. Retrieved December 20, 2023, from https://blog.thirdweb.com/mpc-wallet/

Microsoft. (2018). *Microsoft SEAL: Fast and Easy-to-Use Homomorphic Encryption Library*. Microsoft. Retrieved December 20, 2023, from https://www.microsoft.com/en-us/research/project/microsoft-seal/

Nagaraj, K. (2023, March 7). *Understanding ChaCha20 Encryption: A Secure and Fast Algorithm for Data Protection | 2023*. Karthikeyan Nagaraj. Retrieved December 17, 2023, from https://cyberw1ng.medium.com/understanding-chacha20-encryption-a-secure-and-fast-algorithm-for-data-protection-2023-a80c208c1401

Olier, I. (2018). *A Bayesian approach to identify Bitcoin users*. NCBI. Retrieved December 17, 2023, from https://www.ncbi.nlm.nih.gov/pmc/articles/PMC6292573/

Rasure, E. (2021, October 25). *Zk-SNARK: Definition, How It's Used in Cryptocurrency, and History*. Investopedia. Retrieved December 20, 2023, from https://www.investopedia.com/terms/z/zksnark.asp

Sanak, T. (2019, November 13). *BIP 324 Could Protect Bitcoin Peers*. Bitcoin Magazine. Retrieved December 17, 2023, from https://bitcoinmagazine.com/technical/bip-324-a-message-transport-protocol-that-could-protect-bitcoin-peers

Stanford. (2021). *Distributed Hash Tables with Kademlia—Stanford Code the Change Guides documentation*. Code the Change–Stanford. Retrieved December 17, 2023, from https://code-thechange.stanford.edu/guides/guide_kademlia.html

Vac. (2022, May 17). *Noise handshakes as key-exchange mechanism for Waku*. Vac. Retrieved December 17, 2023, from https://vac.dev/rlog/wakuv2-noise/

Waku. (2023, December 7). *Waku launches first decentralised, privacy-preserving DoS protections for P2P Messaging*. Cointelegraph. Retrieved December 17, 2023, from https://cointelegraph.com/press-releases/waku-launches-first-decentralised-privacy-preserving-dos-protections-for-p2p-messaging

Weinkauf, D. (2023, October 24). *Privacy Tech-Know blog: Computing while blindfolded – Lifting the veil on homomorphic encryption–Office of the Privacy Commissioner of Canada*. The Office of the Privacy Commissioner of Canada's blog. Retrieved December 20, 2023, from https://www.priv.gc.ca/en/blog/20231024/

Wilmoth, J. (2021, March 4). *Bitcoin Privacy Takes Another Step Forward with Wasabi Wallet Launch*. CCN.com. Retrieved December 17, 2023, from https://www.ccn.com/bitcoin-privacy-takes-another-step-forward-with-wasabi-wallet-launch/

Wolford, B. (2022). *What is GDPR, the EU's new data protection law?–GDPR.eu*. GDPR compliance. Retrieved December 20, 2023, from https://gdpr.eu/what-is-gdpr/

Jerry Huang has worked as a technical and security staff at several prominent technology companies, gaining experience in areas like security, AI/ML, and large-scale infrastructure. At Metabase, an open-source business intelligence platform, he contributed features such as private key management and authentication solutions. As a Software Engineer at Glean, a Generative AI search startup, Jerry was one of the three engineers responsible for large-scale GCP infrastructure powering text summarization, autocomplete, and search for over 100,000 enterprise users. Previously at TikTok, Jerry worked to design and build custom RPCs to model access control policies. And at Roblox, he was a Machine Learning/Software Engineering Intern focused on real-time text generation models. He gathered and cleaned a large multilingual corpus that significantly boosted model robustness. Jerry has also conducted extensive security and biometrics research as a Research Assistant at Georgia Tech's Institute for Information Security & Privacy. This resulted in a thesis on privacy-preserving biometric authentication. His academic background includes a BS/MS in Computer Science from Georgia Tech and he is currently pursuing an MS in Applied Mathematics at the University of Chicago. phone: 571–268-6923;

Ken Huang is the author and chief editor of eight books on Generative Artificial Intelligence and Web3, published, respectively, by international publishers including Springer, Cambridge University Press, John Wiley, and China Machine Press. He currently serves as the CEO of the AI and Web3 consulting and education company DistributedApps.AI, based in the United States. Additionally, he holds multiple roles including the expert member of the Blockchain Committee of the Chinese Institute of Electronics, the Co-Chair of AI Organization Responsibility Working Group at Cloud Security Alliance, and Chair of the Blockchain Security Working Group at the Cloud Security Alliance, GCR. He is also a core contributor to the Generative AI Working Group at the NIST and a core author of the OWASP Top 10 for LLM Applications.

Ken Huang has been invited to provide Speaking or Consulting services at institutions including the University of California, Berkeley, Stanford University, Peking University, Tsinghua University, Shanghai Jiao Tong University, China Pacific Insurance, and the World Bank in the past.

Moreover, he has given keynote speeches at international conferences, such as:
– The Davos World Economic Forum 2020 Blockchain Conference
– Consensus 2018 in New York
– The American ACM AI & Blockchain Decentralized Annual Conference 2019
– IEEE Technology and Engineering Management Society Annual Meeting 2019
– Silicon Valley World Digital Currency Forum
– Sino-US Blockchain Summit in Silicon Valley

He has also been awarded the "Blockchain 60" Figure Award by the National University Artificial Intelligence and Big Data Innovation Alliance Blockchain Special Committee in China in 2021.

Mudi Xu is the Executive Deputy Secretary-General of the Cloud Security Alliance Greater China Region and a Ph.D. candidate at the Macau University of Science and Technology, specializing in network security. She has organized and participated in the writing of industry standards such as Cloud Native Security Technical Specifications; and "Cloud Application Security Technical Specifications," as well as over ten white papers, guides, and courses on zero trust security, data security, and cloud security. In China, she planned and published the "Cloud Security Alliance Series," which includes books such as "5G Security: A Compendium of Network Security in the Intelligent Era," Guide to the Field of Data Security, "Zero Trust from Beginner to Expert," and "Zero Trust Network Security: A Guide to Software-Defined Perimeter SDP Architectural Technology."

Part III
Web3 Innovations Are at the Crossroads of Progress and Peril

"With great power comes great responsibility" is an adage that aptly captures the promise and precariousness of emerging innovations. As technological breakthroughs open new frontiers of possibility, they also surface complex questions around ethics, governance, and long-term trajectories. The accelerating pace of human innovation propels us to a crossroads—Will we harness these tools responsibly toward collective progress or invite existential risks through short-sightedness?

The decentralized architecture of Web3 holds potential to shift historical paradigms of data ownership, financial access, creative liberties, and transparent governance. However, hype surrounding this promise often glosses over the challenges in balancing security, rights, and freedoms across these complex socio-technical systems. Like the mythical figure Janus, Web3 stands with two faces gazing toward opposite horizons—optimistic creation versus dystopian destruction. Where we go next depends both on technological progress and collective choices rooted in values.

This concluding part of the book examines the paradoxical nature of innovations through conceptual lenses to make sense of this apparent duality. Chapter 11 summarizes learnings from Web3 security landscapes thus far and presents forward-looking projections. Peering into the future, IoT security issues around identity, authentication, and encryption for rapidly proliferating smart devices and bridges between physical and virtual worlds appear likely to dominate attention even as threats continue evolving across augmented and virtual realities with dangers from tracking and fake visuals that require tailored access controls and protective measures. Meanwhile, the very landscape of Web3 security itself seems destined for continuous evolution in response to novel attack types attempting to exploit decentralization, necessitating agile innovation in threat intelligence, defense in depth, and nimble responsiveness. Also, the integration of AI agents with Web3 requires governance around authorized behaviors. As Web3 evolves, its security frontiers will remain dynamic and intertwined across technologies at this crossroads of progress and peril.

We hope perspectives presented in this book enriched understanding on current issues around Web3, blockchain, AI, and modern technological change while also revisiting timeless dilemmas of the human condition. May this book spur imaginative, thoughtful, and compassionate dialogues around balancing innovation optimally with ethics and positive sum progress. The future remains unwritten, gestating in the collaborative choices made today by people navigating this crossroads one step at a time armed with knowledge. In that spirit, we conclude here inviting you to reflect on learnings from the past and possibilities ahead as participants in shaping what's next for decentralized technology.

Chapter 11
Summary and Future Trends

Jerry Huang ⓘ, **Ken Huang** ⓘ, and **Winston Ma**

Abstract This concluding chapter crystallizes the evolution of Web3 security thus far even as innovations continue apace. Emerging frontiers in quantum cryptography, AI security, IoT, and AR/VR are analyzed to highlight new attack surfaces. Community vigilance through education, responsible innovation, and collaboration is reaffirmed as vital to balance open access and resilience. As Web3 promises a new trust paradigm for digital ecosystems, robust security is highlighted as foundational to mainstream adoption.

The emergence of Web3 introduces a paradigm shift toward a decentralized, user-centric internet, fundamentally built on blockchain, cryptography, and open access protocols. This evolution, while promising, unveils a dynamic and complex cybersecurity landscape. As we navigate through Web3's potential and inherent vulnerabilities, we gather critical insights for future preparedness. This chapter aims to deepen our understanding of Web3's security aspects, highlighting the interplay between foundational security principles and emerging technologies in areas like quantum computing, AI, IoT, and AR/VR. This knowledge is crucial for effectively addressing the unique security challenges presented by these new Web3 frontiers.

J. Huang
The University of Chicago, Chicago, IL, USA
e-mail: jerryh@uchicago.edu

K. Huang (✉)
DistributedApps LLC, Fairfax, Virginia, USA
e-mail: ken@distributedapps.ai

W. Ma
New York University, Broadway, NY, USA

K. Huang et al. (eds.), *Web3 Applications Security and New Security Landscape*,
Future of Business and Finance, https://doi.org/10.1007/978-3-031-58002-4_11

11.1   Recapitulation: Web3 Security Landscapes

A look back offers clarity for the road ahead. This section revisits the pivotal security topics associated with Web3 projects, encapsulating their essence and drawing connections between them.

11.1.1   The Evolving Definition and Components of Web3

The digital landscape has seen a seismic shift with the emergence of Web3, a term that encapsulates the next evolution of the internet. But as with all revolutions, defining its boundaries and understanding its core components has been a journey in itself. In this subsection, we retrace the steps of this evolution, drawing a comprehensive picture of what Web3 has come to signify and the challenges and opportunities it has presented along the way.

Web3 heralds a move from centralized digital platforms, controlled by a few entities, toward a decentralized paradigm where power and control are distributed. At its core, Web3 represents a more open, permissionless, and trustless internet, built upon the bedrock of blockchain technology. It envisions a space where users have true ownership of their data, digital assets, and online identities, a stark departure from the Web2 era where platforms held sway over user data.

But this transformation isn't just philosophical; it's deeply technical. The Web3 era has been marked by the rise of decentralized applications (dApps), smart contracts, and decentralized autonomous organizations (DAOs). These components, powered by blockchain and other decentralized technologies, have redefined how applications are built, how value is transferred, and how online communities and organizations operate.

However, with innovation comes complexity. As we've seen throughout this book, the very features that make Web3 revolutionary—its decentralization, its reliance on consensus mechanisms, its integration of advanced cryptographic techniques—also introduce unique challenges. Issues of scalability, smart contract security, privacy, interoperability between different blockchains, and the integration of legacy systems with new decentralized solutions have been persistent topics of discussion and areas of intensive research.

Moreover, as Web3 has evolved, its definition has expanded. It's no longer just about blockchain. Technologies like privacy-preserving technology, metaverse, AR/VA, AI, Quantum Computing, Decentralized Identity, and advanced cryptographic methods have all come under the Web3 umbrella. Each brings its own set of advantages and challenges, adding layers of complexity to the Web3 narrative.

One of the most profound challenges, as explored in the preceding chapters, is security. The decentralized nature of Web3, while eliminating some traditional vulnerabilities, has introduced new attack vectors. From smart contract vulnerabilities to the challenges of ensuring data privacy in a transparent system, Web3's security landscape is vast and intricate.

In essence, Web3, in its relatively short existence, has transformed the digital realm's fabric. Its definition has expanded and evolved, reflecting the rapid pace of innovation and the challenges that have emerged. As we look back at this journey, it's evident that understanding Web3's multifaceted identity, its core components, and the inherent challenges is crucial. It not only offers a foundation for current operations but also provides a lens to envision the future, ensuring that as we move forward, we do so with clarity, purpose, and an appreciation for the journey thus far.

11.1.2 Threats and Challenges: A Retrospective

The emergence of Web3, with its transformative potential, has also brought to light a labyrinth of security threats and challenges. Each innovation, while pushing the boundaries of digital possibilities, has been matched with a unique set of vulnerabilities. As we reflect on the spectrum of these challenges, it becomes evident that the road to a secure and decentralized internet is intricate, demanding both vigilance and adaptability.

This book provided a comprehensive overview of the key security considerations for various web3 applications including decentralized finance platforms, NFT marketplaces, DAOs, crypto exchanges, and central bank digital currencies. After exploring the security of these applications, part two of this book covered the following advanced security topics:

Quantum Threats: At the intersection of quantum computing and cryptography lies one of the most profound challenges for Web3. Quantum computers, with their ability to process vast amounts of data simultaneously, pose a potential threat to the cryptographic underpinnings of blockchain and other Web3 technologies. Traditional cryptographic techniques, which rely on the difficulty of certain mathematical problems, might be rendered obsolete in the face of a sufficiently powerful quantum computer. The implications are vast—from the integrity of transactions to the security of digital identities.

AI Vulnerabilities: AI's integration into the Web3 ecosystem has been a double-edged sword. While AI-driven security solutions offer proactive threat detection and adaptive responses, the very nature of AI—its reliance on vast datasets, its opaque decision-making processes, and its susceptibility to biases—introduces vulnerabilities. Manipulation of generative outputs, over-reliance on AI solutions, and ethical concerns have emerged as pivotal challenges, demanding a balance between AI's potential and its pitfalls.

Supply Chain Risks: Web3's reliance on third-party libraries, frameworks, and service providers has opened the door to supply chain vulnerabilities. A flaw in a single component can ripple across the ecosystem, underlining the interconnectedness and interdependence of Web3 applications. From vulnerable open-source libraries to service provider downtimes, the supply chain risks are manifold, emphasizing the need for rigorous vetting, monitoring, and diversified dependencies.

Ransomware Evolution: With the digital realm becoming increasingly valuable, ransomware attacks have adapted and evolved. Web3 applications, with their transparent and immutable nature, present both opportunities and challenges in the face of ransomware threats. Understanding the modus operandi of these attacks, from infiltration to ransom demand, has been crucial in crafting robust defense mechanisms.

Privacy and Computation Concerns: The decentralized world of Web3, while offering unparalleled transparency, has also posed challenges for privacy-preserving computations. Techniques like homomorphic encryption and zero knowledge proofs have come to the forefront, offering solutions but also presenting their own set of challenges.

Reflecting on this mosaic of threats and challenges provides invaluable insights. It underscores the dynamism of the Web3 security landscape and the need for continuous adaptation. Each challenge, while posing risks, also offers opportunities—to innovate, to adapt, and to strengthen. As Web3 continues its march toward a decentralized future, understanding these challenges, their implications, and the solutions they've spurred is pivotal. It not only equips us to navigate the present but also to shape a secure, decentralized digital realm for the future.

This book is a companion book to another book titled "A Comprehensive Guide for Web3 Security: From Technology, Economic and Legal Aspects" by the same editors of this book (Huang et al., 2023), readers are highly encouraged to read that book to gain additional insights of Web3 security.

11.1.3 Measures and Solutions: Lessons Learned

In our journey through the intricate terrain of Web3's security landscape, the importance of robust defensive strategies has been a recurrent theme. Facing an array of challenges, the Web3 community has been steadfast in its pursuit of solutions. These measures, a blend of innovation and adaptation, form the pillars of Web3's defense. As we reflect on these strategies, we gain a holistic view of the lessons learned, offering a roadmap for future endeavors.

Proactive Monitoring and Regular Audits: One of the primary lessons has been the significance of proactive measures. Regularly auditing smart contracts, open-source libraries, and other components of the Web3 ecosystem ensures early detection of vulnerabilities. Continuous monitoring, on the other hand, offers real-time threat detection, allowing for immediate mitigation.

Multi-layered Defense Strategies: Relying solely on a single defense mechanism can lead to vulnerabilities. A multi-layered approach, combining AI solutions with traditional security measures, ensures that even if one layer is compromised, others remain intact. This redundancy is crucial for ensuring the resilience of Web3 applications.

Ethical and Transparent AI Deployment: As AI becomes increasingly integrated into the Web3 landscape, ensuring its ethical and transparent deployment is paramount. Frameworks emphasizing explainability, fairness, and bias mitigation ensure that AI operates in alignment with the ethos of decentralization and transparency inherent to Web3.

Quantum-Resilient Cryptography: In anticipation of the quantum computing era, there's been a push toward quantum-resistant cryptographic methods. These techniques, designed to withstand the computational prowess of quantum machines, will be crucial for safeguarding the integrity and security of future Web3 applications.

Diversification and Decentralization: Centralization, even in the decentralized world of Web3, can be a vulnerability. Diversifying service providers, data storage solutions, and even consensus mechanisms ensures that there's no single point of failure, enhancing the resilience of the ecosystem.

Education and Training: At the heart of many vulnerabilities lies human error. Regular training sessions, workshops, and educational resources ensure that developers, stakeholders, and users are aware of best practices, emerging threats, and the latest countermeasures.

Community Collaboration: Web3's strength lies in its community. Collaborative efforts, where developers, researchers, and users come together, have been instrumental in identifying vulnerabilities, crafting solutions, and sharing knowledge. Open-source collaborations, community-driven audits, and decentralized governance models exemplify the power of collective effort in securing the Web3 space.

In retrospect, the journey through Web3's security challenges and solutions offers a tapestry of lessons. The threats, while formidable, have been matched with ingenuity, collaboration, and a relentless pursuit of security. These lessons, distilled from myriad experiences and challenges, form a foundation upon which the future of Web3 security will be built. They serve as a testament to the community's resilience, adaptability, and commitment to creating a secure, decentralized digital future.

11.2 Peering into the Future: Emerging Security Frontiers

While the present holds its challenges, the future beckons with new frontiers. This section delves into imminent areas of interest, underscoring the security considerations they bring forth.

11.2.1 IoT Security: The Next Battleground

The Internet of Things (IoT) represents a vision where everyday objects, from refrigerators to vehicles, are embedded with sensors, software, and other technologies to connect and exchange data with other devices and systems over the internet.

The promise of IoT is immense, ushering in smart cities, optimized industries, and enhanced personal experiences. However, this interconnected ecosystem also presents a plethora of security challenges, especially when viewed in the context of Web3's decentralized ethos.

As the boundaries between the physical and digital realms blur with IoT, the surface for potential attacks expands. Each device, each connection becomes a potential entry point for malicious actors. The stakes are high. A vulnerability in a smart home system can compromise personal privacy, while a breach in an industrial IoT setup can have catastrophic repercussions on production and safety.

Web3's intersection with IoT further amplifies these challenges. Imagine a decentralized application (dApp) that manages a smart grid system or an IoT-based supply chain. While Web3 offers transparency and decentralization, ensuring the security of countless devices in such a network becomes paramount. Smart contracts, for instance, could control IoT devices based on certain triggers or data inputs. A flaw in such a contract could thus have real-world consequences, making the validation and security of these contracts even more critical.

Moreover, the heterogeneity of IoT devices—varying manufacturers, diverse operating systems, and different software versions—complicates standardization and security protocol implementation. As Web3 platforms interact with these varied devices, ensuring consistent security practices becomes a monumental task.

Data privacy is another significant concern. IoT devices constantly gather data, and when stored on a transparent and immutable blockchain, ensuring that this data doesn't infringe on personal privacy rights is crucial. Techniques like zero-knowledge proofs and other privacy-preserving methods discussed earlier in the book become even more relevant in this context.

To navigate this intricate landscape, several measures can be anticipated:

11.2.1.1 Device Identity and Authentication

Establishing a decentralized identity for Internet of Things (IoT) devices within a Web3 framework can enhance network security. This approach ensures that only authenticated and verified devices can interact with the network, offering a more secure and autonomous way to manage device identities compared to traditional centralized systems.

In the context of Web3, decentralized identity allows each IoT device to have a unique, cryptographic identity, typically managed through blockchain technology. This technology provides an immutable record of the device's identity and transactions, with transparency and tamper-resistance ensuring secure and verifiable identities for each device.

Authentication in this setup involves validating the device's identity before it can communicate or perform actions within the network. This is often achieved using cryptographic techniques such as digital signatures, where a device signs its communications with a private key, and the corresponding public key is used by other devices and network participants to verify the signature and confirm the device's authenticity.

The benefits of decentralized identity and authentication for IoT devices in a Web3 environment are significant. It reduces the risk of identity spoofing and fraudulent activities, as cryptographic proof of identity is much harder to compromise compared to traditional username-password systems. It also enables a more scalable and efficient way to manage identities, as the blockchain can handle numerous transactions and identity verifications without a central authority. Furthermore, it promotes interoperability and easier integration of devices across different networks and platforms, assuming the decentralized identity standard is universally recognized and adopted.

Implementing this concept involves several key steps, including designing a secure and resilient identity structure on the blockchain, equipping devices with the necessary hardware with trusted execution environment and software capabilities to generate and manage cryptographic keys in a secured enclave, and establishing robust and efficient protocols for identity verification and authentication.

Incorporating decentralized identity and authentication for IoT devices in a Web3 framework is a forward-thinking approach that enhances network security, streamlines identity management, and improves interoperability across various platforms and devices. This method offers a more secure and autonomous system, which is increasingly important as IoT devices become more prevalent in various applications.

11.2.1.2 End-to-End Encryption

End-to-end encryption is a fundamental aspect of securing communication between Internet of Things (IoT) devices and Web3 platforms. It plays an essential role in maintaining data integrity and confidentiality. By implementing this form of encryption, the data transmitted from IoT devices to Web3 platforms is converted into a secure format that is nearly impossible to decipher by unauthorized parties. This ensures that sensitive information remains confidential and protected from potential breaches or unauthorized access.

In a typical end-to-end encryption scenario within a Web3 and IoT context, data is encrypted at the source, which is the IoT device itself. This encryption process involves using a cryptographic key known only to the device and the intended recipient, which in this case is the Web3 platform. As the data travels across various networks or channels, it remains in its encrypted form, making it unreadable to anyone who might intercept it. The data is only decrypted when it reaches the intended recipient, ensuring that only authorized parties can access the original information.

The importance of end-to-end encryption in this setting cannot be overstated. IoT devices often collect and transmit sensitive data, including personal user data, operational data, or critical infrastructure information. Without robust encryption, this data could be vulnerable to interception, leading to privacy breaches, data manipulation, or other malicious activities. Furthermore, in a Web3 environment, where decentralized applications and smart contracts might automatically act on this data, ensuring its integrity is paramount. Any tampering or alteration of the data could lead to incorrect operations or transactions.

Implementing end-to-end encryption in IoT and Web3 systems requires careful consideration of the encryption methods and standards. Advanced encryption algorithms like AES (Advanced Encryption Standard) are commonly used due to their reliability and security. Additionally, the management of cryptographic keys is vital. Secure key generation, distribution, and storage mechanisms must be in place to prevent unauthorized access to the keys. NIST guidance on key management can be leveraged for this purpose (NIST, 2020).

Moreover, the integration of end-to-end encryption must consider the constraints of IoT devices. These devices often have limited processing power and energy resources, making it necessary to implement encryption methods that are efficient yet secure. Lightweight cryptographic algorithms are particularly suitable for such environments (Delvadiya, 2023).

In conclusion, end-to-end encryption is a vital component in the secure communication between IoT devices and Web3 platforms. It ensures that data remains confidential and unaltered during transit, safeguarding against unauthorized access and data breaches. The implementation of robust encryption practices, combined with efficient key management and consideration for the limitations of IoT devices, is essential in maintaining the security and integrity of data in these interconnected systems.

11.2.1.3 Regular Device Updates

Regular device updates play a role in maintaining the security and functionality of Internet of Things (IoT) devices, especially when they are integrated with Web3 platforms, similar to the ongoing audits and updates required for smart contracts and decentralized applications (dApps). These updates are not just routine maintenance tasks; they are essential for patching known vulnerabilities, enhancing features, and ensuring the overall resilience of the IoT ecosystem within the dynamic landscape of Web3 technologies.

IoT devices, by their nature, are often deployed in various environments, collecting and processing data, and interacting with other devices and systems. Over time, vulnerabilities may be discovered in their firmware or software. These vulnerabilities can be exploited by attackers to gain unauthorized access, disrupt device functionality, or compromise data integrity. Regular updates help in mitigating these risks by patching security holes and reinforcing the device's defenses against emerging threats.

In the context of Web3, where IoT devices may interact with blockchain-based platforms, the need for regular updates is even more pronounced. The decentralized nature of Web3 applications means that security and functionality issues can have far-reaching consequences. For example, an unpatched vulnerability in an IoT device could potentially lead to the compromise of sensitive data or manipulation of smart contract interactions via decentralized data feeder. Therefore, keeping IoT devices up-to-date is crucial for maintaining the trust and reliability of the entire system.

Implementing a robust update mechanism for IoT devices involves several considerations:

1. Secure Update Process: The update process itself must be secure to prevent the introduction of malware or unauthorized modifications. This often involves cryptographic signing of updates and secure transmission channels.
2. Minimal Disruption: Updates should be designed to cause minimal disruption to the device's operation, especially for critical applications where downtime can have significant implications.
3. Compatibility and Testing: Ensuring that updates do not negatively impact the device's compatibility with other systems or its core functionalities is essential. Rigorous testing before deployment can mitigate risks associated with updates.
4. Automated Update Mechanisms: Automated or semi-automated update mechanisms can ensure timely application of patches and reduce the reliance on manual processes, which can be prone to delays or errors.
5. User Notification and Control: For certain types of updates, especially those that might change the functionality or privacy settings of a device, notifying users and giving them some control over the update process can be important.

Regular updates for IoT devices, particularly in a Web3 setting, are not just about fixing bugs or closing security gaps. They also provide opportunities to enhance device capabilities, optimize performance, and ensure compatibility with evolving standards and technologies in the rapidly changing digital landscape. A proactive and well-managed update strategy is therefore essential for maintaining the security, functionality, and longevity of IoT devices in a Web3 ecosystem.

11.2.2 Immersive Realities: AR/VR/XR/MR Security

The dawn of immersive realities—Augmented Reality (AR), Virtual Reality (VR), Mixed Reality (MR), and Extended Reality (XR)—heralds a new era of digital experiences. These technologies, by superimposing digital content onto our physical world or creating entirely virtual spaces, offer transformative potential in fields ranging from entertainment and gaming to healthcare and education. Yet, as we dive deeper into these immersive worlds, we're confronted with a suite of unique security challenges when these realities intersect with the decentralized tenets of Web3.

11.2.2.1 Potential AR/VR/XR/MR Security Issues

The following are some security issues associated with AR/VR/XR/MR.

Data Privacy and Personal Information: Immersive realities often rely on a slew of sensors to function—cameras, microphones, and even biometric sensors. This means they collect vast amounts of data, some of it deeply personal. In a VR social space, for instance, user behaviors, interactions, even facial expressions could be recorded and analyzed. For example, Rutgers University researchers developed an attack called "Face-Mic" that can steal information from popular VR headsets. The attack works by analyzing vibrations from the headset's motion sensors. These vibrations can reveal a user's credit card numbers, passwords, and other sensitive information. The researchers hope that their findings will encourage manufacturers to develop more secure VR headsets (Chen, 2022). When AR/VR data is stored or processed on Web3 platforms, ensuring its privacy becomes paramount. A breach here could reveal intimate details, not just digital credentials.

Malicious Digital Overlays in AR: Augmented Reality overlays digital content onto the physical world. Imagine malicious actors manipulating this to mislead users—redirecting them to hazardous locations or overlaying false information on genuine physical objects. The consequences range from misinformation to physical harm.

For example, the research papers describe how a team of research devised attacks capable of potentially disorienting users, activating their Head Mounted Display (HMD) camera surreptitiously, superimposing images within their field of vision, and manipulating VR environmental factors to compel inadvertent collisions with physical objects and walls. Ultimately, the study demonstrates, through a deception experiment involving human participants, the effectiveness of exploiting AR/VR systems to manipulate immersed users and guide them to a location in physical space without their awareness. This phenomenon is termed the Human Joystick Attack. The research concludes by outlining future directions for exploration and suggesting measures to bolster the security of these systems (Casey et al., 2019).

As another example, there was malware called Vizom that uses remote overlay attacks to hijack bank accounts. It targets users in Brazil by disguising itself as popular video conferencing software. Once installed, it steals banking credentials and can even take control of a user's computer. The malware is spread through phishing emails and can infect any Windows PC. IBM researchers believe that the same tactics are being used to target users in other parts of South America and Europe (Osborne, 2020).

Digital Identity and Deepfakes in VR: VR spaces allow users to craft avatars and digital personas. The threat of deepfakes—hyper-realistic but entirely fake content—enters this space, allowing malicious actors to impersonate others, spread false information, or manipulate perceptions. For example, fraudsters are using AI and deepfakes to commit identity fraud (Margalit & Kempf, 2023) by creating fake IDs and videos to trick facial recognition systems in VR systems.

Hardware Vulnerabilities: The devices powering these immersive experiences, from VR headsets to AR glasses, can be targeted. Malware injected into these devices could distort realities, spy on users, or even cause physical discomfort. For example, Malware such as "Big Brother" has been identified as a threat to VR devices, allowing attackers to remotely record the user's headset screen, potentially leading to privacy breaches and corporate espionage (Kelly, 2022).

Economic Implications in Virtual Spaces: As Web3 enables virtual economies in these digital spaces—think virtual real estate or digital art galleries—there are economic implications to security breaches. Theft, fraud, and economic manipulations become real concerns.

11.2.2.2 Some Mitigation Strategies

Navigating these challenges requires some risk mitigation strategies and actions:

1. Robust Authentication Mechanisms: Given the intimate nature of data collected, multi-factor authentication, possibly even leveraging biometrics, can ensure only authorized access to devices and digital spaces.
2. Data Minimization and Encryption: Only collect essential data and ensure it's encrypted both in transit and at rest. Decentralized storage solutions could further ensure data isn't susceptible to centralized breaches.
3. Digital Etiquettes and Norms: As with any society, virtual spaces will require etiquettes and norms. Clear guidelines on acceptable behaviors, combined with mechanisms to report and act against violations, will be crucial.
4. Continuous Device Updates: Regular updates to the firmware and software of devices can patch known vulnerabilities, ensuring malicious actors can't exploit them.

In essence, while the universes of AR, VR, MR, and XR beckon with unparalleled experiences, they also usher in a new frontier of security challenges. Balancing the promise of these technologies with their potential pitfalls is the next great challenge, demanding both innovation and vigilance. Issues around fraud, hacking, identity theft, and more could easily undermine user trust and inhibit mainstream adoption of Web3-connected XR if left unaddressed. As virtual and augmented worlds become more ingrained into work, social, financial, and daily life, security attacks could have material consequences well beyond just the digital domain.

11.2.3 The Continuous Evolution of Web3 Security

The realm of Web3, by its very nature, is poised for continuous evolution. Just as the internet of the 1990s is a far cry from today's digital expanse, Web3 today might seem rudimentary a decade from now. This dynamism inherently means that the security landscape of Web3 will be in perpetual flux, facing new threats while also harnessing innovations to combat them.

11.2.3.1 New Security Landscapes

Emerging Technologies and New Threat Vectors: The integration of newer technologies into the Web3 framework will invariably introduce fresh vulnerabilities. For instance, as quantum computing inches closer to practicality, its potential to disrupt traditional cryptographic methods will be a looming challenge for Web3 platforms. Similarly, as we integrate brain computer interfaces or other futuristic technologies, the definition of "personal data" and its potential misuse will expand dramatically.

Inter-Blockchain Communication and Multi-chain Challenges: As multiple blockchains become interoperable, allowing assets and data to flow between different chains, the complexity of ensuring security multiplies. Each chain might have its own vulnerabilities, and ensuring consistent security practices across chains becomes paramount.

Decentralized Identities and Personal Sovereignty: As Web3 moves toward truly decentralized identities, where individuals have complete control and ownership of their digital selves, the challenges of ensuring security without centralized oversight become pronounced. Phishing attacks, identity thefts, or data losses could have severe repercussions in a world where digital identity is sacrosanct.

Evolving Regulatory Landscapes: As governments and institutions grapple with the decentralized ethos of Web3, regulatory frameworks will evolve, sometimes in reactionary ways. Navigating these changing regulations, ensuring compliance while also safeguarding user rights and privacy, will be a continuous challenge. The recent Binance legal case highlights the importance of proactive regulatory compliance to avoid hefty fines (Michaels, 2023).

Yet, with challenges come innovations:

11.2.3.2 Some New Security Innovations

1. Adaptive Cryptography or crypto-agility: Future cryptographic methods will likely be adaptive, capable of morphing in response to the computational capabilities of potential threats (McCarthy, 2023). Quantum-resistant algorithms are just the tip of the iceberg in this evolution.
2. Decentralized AI for Security: As AI models become more decentralized, their application in real-time threat detection and response across Web3 platforms could be a game changer. These models, trained continuously on network data, can preemptively detect and neutralize threats.
3. Community-driven Security Initiatives: The strength of Web3 lies in its community. Crowdsourced security solutions, decentralized audits, and community-driven threat intelligence will play pivotal roles in safeguarding the ecosystem. For example, community-driven web3 bug bounty program Immunefi played a good role in this space (Butcher, 2022).

We can expect that the journey ahead for Web3 security is both exhilarating and daunting. The challenges will be multifaceted, but so will the solutions. The community, armed with lessons from the past and innovations of the present, will forge ahead, ensuring that the decentralized promise of Web3 remains both secure and inclusive. The road might be winding, but the destination—a secure, decentralized digital realm—is worth the journey.

11.2.4 AI Agent with Web3

AI Agent (or Assistant) powered by a Large Language Model can one day have the planning, reflection, self-correction, and profound reasoning capabilities that can be used in Web3 ecosystems. In this section, we will look at the possibility of such an agent and how it can be used for Web3 and what are security implications.

11.2.4.1 AI Agent for Web3

Imagine an AI assistant that not only executes tasks but also plans its actions based on an understanding of the blockchain environment. This planning capability becomes useful in managing smart contracts and decentralized applications (DApps). The AI could forecast potential issues in smart contract execution or network congestion, proactively suggesting adjustments to optimize performance and cost. In DApp development, such foresight could lead to more user-friendly interfaces and efficient resource allocation, dynamically adapting to user behaviors and network conditions.

The reflection aspect of AI can be transformative in analyzing blockchain transactions and trends. By reflecting on past events and outcomes, AI can provide insights into market trends and user behaviors, offering valuable predictions and strategies for traders and developers alike. This reflective analysis could also extend to security, where the AI examines past security breaches or vulnerabilities across the blockchain, learning from them to bolster the security of current systems.

Self-correction is another vital facet. In the Web3 space, where the stakes are high due to the immutable nature of blockchain transactions, an AI's ability to self-correct in real-time can prevent costly errors. For instance, in automated trading or decentralized finance (DeFi) protocols, an AI that can quickly rectify its strategies in response to sudden market shifts could protect investments from significant losses. Similarly, in smart contract execution, an AI that self-corrects coding errors or vulnerabilities before they are deployed on the blockchain could save developers from irreversible consequences.

The profound reasoning capability of AI can be the cornerstone of innovative solutions in Web3. This capability enables AI to not just analyze data but to understand and infer complex patterns and relationships within the blockchain ecosystem. Such reasoning could lead to the development of more sophisticated and secure DeFi products, predict and adapt to regulatory changes, and even contribute to more sustainable blockchain operations by optimizing energy usage patterns in mining or validating processes.

Furthermore, these AI capabilities can revolutionize user interaction within the Web3 space. From personalized investment advice in DeFi to tailored experiences in DApps, AI can use its reasoning and reflection abilities to understand and predict user needs, enhancing user satisfaction and engagement.

11.2.4.2 Security Implication of AI Agent in Web3

The integration of AI assistants or agents with advanced capabilities like planning, reflection, self-correction, and profound reasoning into Web3 ecosystems can significantly enhance security, but it also introduces unique challenges that require careful consideration. These AI capabilities, when applied thoughtfully, can fortify the security framework of Web3, yet their complexity and sophistication necessitate a nuanced approach to ensure they do not inadvertently introduce new vulnerabilities.

Firstly, the planning ability of AI can be a double-edged sword. While it enables proactive identification and mitigation of potential security threats in the blockchain network, it also raises concerns about the AI's decision-making process. The AI's plans must be transparent and interpretable to ensure that they align with security protocols and do not unintentionally compromise the system's integrity. For instance, an AI planning to optimize network traffic for a DApp must ensure that its strategies do not inadvertently create openings for DDoS attacks or expose sensitive data.

Reflection and learning from historical data is invaluable for security, especially in identifying and understanding past vulnerabilities and attacks. However, the reflection process must be meticulously managed to avoid biases and errors in learning. An AI reflecting on incomplete or biased data could develop flawed understandings of security threats, leading to inadequate or inappropriate security measures.

The self-correction capability of AI is particularly critical in real-time security scenarios. It allows the AI to adapt and respond quickly to emerging threats, such as unusual transaction patterns that might indicate a breach or an ongoing attack. However, this self-correction should be balanced with human oversight to prevent over-correction or the AI making drastic changes that could destabilize the system.

Profound reasoning allows AI to understand complex and abstract concepts, which can be a significant advantage in predicting and countering sophisticated cyber threats. This capability enables AI to analyze the intentions behind transactions or interactions within the blockchain, potentially identifying malicious actors or complex fraud schemes. However, the depth of reasoning must be carefully calibrated to respect privacy and avoid misinterpreting legitimate activities as threats.

Moreover, the integration of AI into Web3 must consider the immutable nature of blockchain. Decisions or actions taken by AI, especially those that interact directly with the blockchain, must be reversible or adjustable in some form, as the blockchain itself does not allow for easy correction of errors once recorded.

Lastly, as AI systems become more integral to the security of Web3 ecosystems, they themselves become targets for attackers. Ensuring the security of the AI—protecting its data sources, learning algorithms, and operational integrity—becomes paramount. This includes guarding against adversarial attacks designed to deceive or manipulate the AI's learning process or output.

11.3   Concluding Thoughts and Way Forward

We will share our reflections on the journey traversed, emphasizing the importance of vigilance, innovation, and collaboration in navigating the future of Web3 security.

11.3.1   The Imperative of Collaborative Security

The decentralized nature of Web3 is both its strength and its vulnerability. It lacks a central authority, making decisions and implementing changes a collective endeavor. In such a landscape, the role of collaborative security becomes not just desirable but imperative.

Every stakeholder, from individual users and developers to organizations and regulators, holds a piece of the security puzzle. Gone are the days when security was the domain of a select few experts. In the world of Web3, security is everyone's concern. This democratization of responsibility offers both challenges and opportunities.

When a vulnerability is identified in a decentralized application, it's not just the developers who rally to address it. The community at large, equipped with diverse skills and perspectives, often converges to dissect the problem, propose solutions, and implement fixes. This hive mind approach means that solutions are not just rapid but also holistic, taking into account a multitude of perspectives.

Moreover, as Web3 platforms and applications proliferate, the interdependence between them grows. A security flaw in one platform might have ripple effects across the ecosystem. Recognizing this interconnectedness, platforms have started collaborating, sharing threat intelligence, and jointly developing mitigation strategies.

But collaboration isn't just about addressing threats; it's also about proactive defense. Joint initiatives to educate users, developers, and organizations about best practices, potential threats, and safe behaviors are sprouting across the Web3 world. These educational endeavors, often driven by community volunteers, highlight the spirit of collective guardianship.

In essence, the decentralized ethos of Web3 extends to its security paradigm. It's a world where collaboration isn't just beneficial—it's essential. By fostering a culture of shared responsibility, transparency, and mutual aid, the Web3 community ensures that the very principles that gave birth to this revolution also safeguard its future.

11.3.2   Education and Awareness: The First Line of Defense

At the intersection of innovation and security, education and awareness play an important role. While state-of-the-art cryptographic techniques and cutting-edge defense mechanisms are undeniably crucial, an informed and vigilant community amplifies their effectiveness manifold.

The Web3 ecosystem, by design, shifts power to individuals, be they users transacting on a decentralized platform or developers crafting the next big dApp. This democratization, while empowering, also places the onus of security on individuals. And the most formidable shield they can wield against potential threats is a deep understanding of the ecosystem and its vulnerabilities.

11.3.2.1   User Education

For the average user, Web3 can be a daunting realm, replete with novel concepts like cryptographic wallets, gas fees, and smart contracts. Without a basic understanding of these elements, users become easy prey for scams, phishing attempts, and other malicious activities. Comprehensive education campaigns, ranging from simple explainer videos to in-depth workshops, can equip users with the knowledge to navigate Web3 safely. By understanding the basics of private key management, recognizing the hallmarks of phishing sites, or being aware of common scams, users can dramatically reduce their risk exposure.

While innovative solutions strive to present a user-friendly interface, simplifying complex blockchain and smart contract concepts, it is crucial that users are not entirely shielded from these underlying mechanisms.

This necessity stems from the unique security challenges and responsibilities inherent in Web3. Unlike traditional web environments, where platforms often assume significant responsibility for user security, Web3 places much of this onus on the individual. Moreover, comprehending the principles of smart contracts and their execution can help users recognize and avoid interacting with malicious contracts designed to siphon funds or personal data.

Awareness of gas fees and their fluctuations is also crucial in making informed decisions about transaction timings and avoiding unnecessary expenses. Educational initiatives, therefore, should focus on imparting this essential knowledge. This could range from creating intuitive tutorials and guides that explain the basics of blockchain and cryptography, to more advanced workshops that delve into smart contract security and risk management strategies in Web3 environments.

By empowering users with this knowledge, they become active participants in their security, capable of identifying and avoiding common scams like phishing attempts, which are prevalent in the Web3 space.

11.3.2.2 Developer Awareness

To ensure that developers remain at the forefront of security in Web3, continuous education and exposure to the latest developments in the field are indispensable. Regularly organized workshops, hackathons, and seminars serve as vital platforms for this purpose. These events provide developers with opportunities to stay abreast of the newest vulnerabilities discovered in blockchain technologies and smart contracts. They also facilitate the sharing of best coding practices and the exploration of emerging defense techniques. Such collaborative and educational environments are essential for fostering a community of developers who are well-informed and skilled in creating secure Web3 applications.

Moreover, these events often simulate real-world scenarios, allowing developers to test and hone their skills in a controlled environment. Hackathons, for instance, can challenge developers to identify and fix vulnerabilities in smart contracts or to build applications with security as a primary focus. Seminars led by experts in blockchain security can offer insights into advanced cryptographic techniques or strategies to mitigate specific threats like front-running or reentrancy attacks.

A developer who is cognizant of the potential attack vectors and the latest security trends is far better equipped to build inherently secure applications. This awareness is not just about preventing attacks but also about instilling trust in the users of Web3 platforms. When developers prioritize security in their design and code, they contribute to a more robust and resilient Web3 ecosystem. This proactive approach to security is vital in an environment where a single vulnerability can have far-reaching and irreversible consequences.

11.3.2.3 Organizational Initiatives

Organizations, particularly those spearheading Web3 initiatives, occupy an essential role in cultivating a culture of security awareness within the ecosystem. Their influence extends beyond their internal operations to the broader community of users, developers, and stakeholders in the Web3 space. By undertaking targeted and strategic initiatives, these organizations can significantly enhance the collective understanding and application of security principles in Web3.

One of the key methods through which organizations can contribute is by hosting regular training sessions. These sessions can be tailored to different levels of expertise, catering to both novice and experienced participants. For newcomers, training might focus on the fundamentals of blockchain technology, smart contract security, and basic cyber hygiene practices. For more advanced users and developers, the content can delve into complex topics like advanced cryptographic techniques, in-depth analysis of smart contract vulnerabilities, and strategies for securing decentralized applications (dApps). These training sessions serve not only as educational platforms but also as opportunities for participants to engage with experts and peers, fostering a collaborative learning environment.

Another significant avenue for organizations to impact security awareness is through the creation and dissemination of user-friendly documentation.

Well-crafted documentation that clearly explains the workings of a Web3 project, including its security features and best practices for users, is invaluable. This documentation can take various forms, such as detailed guides, FAQs, and interactive tutorials, making complex concepts more accessible and understandable to a broader audience.

Active participation in community forums is also a crucial aspect of organizational initiatives in Web3 security education. By engaging in discussions, answering queries, and sharing insights on platforms like Reddit, Discord, and specialized blockchain forums, organizations can directly interact with the community. This engagement not only helps in spreading awareness but also allows organizations to receive feedback and insights from the community, which can be instrumental in improving their projects and security measures.

Moreover, organizations can lead by example, showcasing best practices in their own operations and development processes. This includes conducting regular security audits, being transparent about vulnerabilities and how they are addressed, and adopting a proactive stance in dealing with emerging threats and challenges. Such practices not only enhance the security of their own projects but also set a benchmark for others in the industry to follow.

11.3.2.4 Collaborations

Collaborations between various stakeholders, such as academic institutions, technology companies, and non-profit organizations, stand as a cornerstone in this endeavor. By uniting their diverse resources, knowledge, and outreach capabilities, these entities can significantly magnify the reach and efficacy of educational initiatives in the Web3 space.

Academic institutions bring a wealth of theoretical knowledge and research capabilities to the table. They can delve deep into the foundational aspects of blockchain technology, cryptography, and the principles governing decentralized systems. Their involvement ensures that the educational content is not only current but also grounded in rigorous academic research. Furthermore, universities and research institutions can foster a new generation of Web3 experts through specialized courses, degree programs, and research opportunities.

Technology companies, especially those operating within the Web3 and blockchain domain, offer practical insights and real-world applications of these technologies. Their expertise is invaluable in demonstrating the practical implementation of theoretical concepts. They can provide case studies, best practices, and insights into the latest technological advancements and challenges in the field. These companies can also contribute technological tools and platforms to facilitate learning, such as interactive coding environments, simulation platforms, and access to blockchain networks for experimental purposes.

Non-profit organizations, on the other hand, play a crucial role in ensuring the accessibility and inclusivity of these educational campaigns. They can work toward breaking down complex concepts into more digestible content for the general public, making it easier for novices to grasp the essentials of Web3. Additionally, non-profits can focus on reaching underrepresented groups, ensuring that the benefits and knowledge of Web3 technology are distributed equitably.

When these diverse entities collaborate, they can create comprehensive and multifaceted educational content that caters to a wide range of audiences, from beginners to seasoned professionals. This collaboration can take various forms, such as joint webinars, collaborative research projects, community workshops, and online courses. Such initiatives ensure that the content is not only technically sound but also widely accessible and engaging.

Moreover, these collaborative efforts can lead to the creation of standardized educational frameworks and certifications, which can help in establishing a baseline of knowledge and expertise in the Web3 field. This standardization is particularly important in a field that is as dynamic and rapidly evolving as Web3, where staying updated with the latest developments and security practices is crucial.

11.3.3 A Call to Action: The Road Ahead

Web3 is not just a technological revolution; it's a societal one. It's a reimagining of how we interact, transact, and trust in the digital realm. And with this transformation, every individual, whether a casual user or a seasoned developer, becomes a guardian of this new world. The responsibility of safeguarding Web3 doesn't rest on the shoulders of a select few; it's a collective endeavor, and every contribution, however small, strengthens the fabric of this burgeoning ecosystem.

To the readers, our fellow pioneers of this brave new world, the call to action is clear:

Responsible Innovation: The dynamics of Web3 are in constant flux. Responsible innovation pushes the ecosystem forward while adhering to regulations and applying security controls to Web3 applications.

Engage and Collaborate: Web3's strength lies in its community. Engage with peers, participate in forums, attend workshops, and contribute to open-source projects. The synergy of collective intelligence is what will propel Web3 to its envisioned potential.

Practice Vigilance: Security isn't a one-time effort but a continuous practice. Regularly update your software, back up your data, and be wary of too good to be true offers. Remember, in the decentralized world, your security is in your hands.

Contribute to the Cause: If you possess expertise, consider mentoring others, conducting awareness sessions, or contributing to educational resources. If you're a developer, consider volunteering for security audits or developing tools to enhance Web3 security.

Champion Ethics and Integrity: Beyond technicalities, Web3 is also about ethics, trust, and community. Champion transparency, fairness, and integrity in all your digital interactions. Build and support projects that prioritize user rights, privacy, and security.

As we journey forward, let us remember that Web3 is more than just technology; it's a vision—a vision of a decentralized, inclusive, and secure digital realm. And to realize this vision, every hand, every mind, and every voice counts. Let's embrace the future with optimism, determination, and an unwavering commitment to securing our collective digital destiny. The road ahead is long, but together, we can ensure it leads to a brighter, safer, and more inclusive tomorrow.

11.4   Concluding Thoughts

This book extends our research reported in our previous book titled <A Comprehensive Guide for Web3 Security: From Technology, Economic and Legal Aspects> (Huang et al., 2023). Readers interested in Web3 security are strongly encouraged to read our previous book.

This concluding chapter provided a retrospection of the evolution of Web3 security while peering into imminent frontiers. By recapitulating pivotal concepts, threats, and solutions, it crystallized the intricate interplay between innovation and security in this burgeoning landscape.

Emergent technologies like IoT, immersive realities, and AI assistants usher profound potential but also unique attack surfaces demanding renewed approaches. As quantum computing inches toward viability, the imperative for quantum-resilient cryptography and adaptive techniques heightens. Visionary possibilities like AI agents with planning and reasoning capabilities promise to transform applications but introduce concerns around transparency, oversight, and blockchain irreversibility.

The chapter emphasizes that the Web3 security paradigm must perpetually evolve in response to technological leaps. Innovations around decentralized AI, crowdsourced solutions, interoperable identity management, and more are already emerging to match new threats. However, beyond novel tools and robust cryptography, security awareness through education, responsible disclosure, and community vigilance remains foundational. Users, developers, and enterprises alike need appropriate guidance to foster secure behaviors, applications, and policies, respectively.

Collaboration is affirmed as the cornerstone for security in the decentralized landscape, be it through coordination of disclosure processes or unified patching of exploited vulnerabilities. Complacency has no room when each linkage introduces risks but together, the ecosystem can thrive through shared intelligence. As innovations multiply attack vectors, the collective defense marshaled through crowdsourced testing, standardized guidelines, and rotational multi-layered approaches promises resilience.

The chapter concludes with a call for proactive stakeholder participation through active collaboration, vigilance, transparent development, and contributing within one's capacity to further the vision of a trusted Web3 ecosystem. It re-envisions security as a shared responsibility in this user-powered paradigm. Stressing ethical technology development, it argues that beyond robust privacy and cryptography, upholding principles of inclusivity, transparency, and fair access is vital for mainstream adoption. As Web3 promises to redefine digital trust, getting security right with foresight and wisdom is key to unlocking that revolutionary potential.

References

Butcher, M. (2022, September 22). *Web3 bug-bounty platform Immunefi raises $24M for its Series A funding round*. TechCrunch. Retrieved December 22, 2023, from https://techcrunch.com/2022/09/22/web3-bug-bounty-platform-immunefi-raises-24m-for-its-series-a-funding-round/

Casey, P., Baggili, I., & Yarramreddy, A. (2019). *Immersive Virtual Reality Attacks and the Human Joystick | Request PDF*. ResearchGate. Retrieved December 22, 2023, from https://www.researchgate.net/publication/332050743_Immersive_Virtual_Reality_Attacks_and_the_Human_Joystick

Chen, J. (2022, February 10). *Rutgers Researchers Discover Security Vulnerabilities in Virtual Reality Headsets*. Rutgers University. Retrieved December 22, 2023, from https://www.rutgers.edu/news/rutgers-researchers-discover-security-vulnerabilities-virtual-reality-headsets

Delvadiya, D. (2023, February 20). *How is Lightweight Cryptography Applicable to Various IoT Devices? Tutorialspoint*. Retrieved December 22, 2023, from https://www.tutorialspoint.com/how-is-lightweight-cryptography-applicable-to-various-iot-devices

Huang, K., Budorin, D., Tan, L. J., Ma, W., & Zhang, Z. W. (Eds.). (2023). *A Comprehensive Guide for Web3 Security: From Technology, Economic and Legal Aspects*. Springer Nature Switzerland.

Kelly, E. (2022, July 25). *Hackers can see what you're doing in VR via Big Brother malware*. VentureBeat. Retrieved December 22, 2023, from https://venturebeat.com/games/hackers-can-see-what-you-are-doing-in-vr-via-big-brother-malware/

Margalit, N., & Kempf, R. (2023, June 26). *How Fraudsters Leverage AI and Deepfakes for Identity Fraud. Transmit Security*. Retrieved December 22, 2023, from https://transmitsecurity.com/blog/how-fraudsters-leverage-ai-and-deepfakes-for-identity-fraud

McCarthy, S. (2023, August 2). The Power of Crypto-Agility*: A Defence-in-Depth Strategy for Quantum and Beyond*. evolutionQ. Retrieved December 22, 2023, from https://www.evolutionq.com/post/crypto-agility-for-quantum-and-beyond

Michaels, D. (2023, December 4). *Binance Copped a $4 Billion Plea but Is Still Fighting the SEC*. The Wall Street Journal. Retrieved December 22, 2023, from https://www.wsj.com/finance/regulation/binance-copped-a-4-billion-plea-but-is-still-fighting-the-sec-44a4e5a5

NIST. (2020, May 4). *NIST Special Publication (SP) 800-57 Part 1 Rev. 5, Recommendation for Key Management: Part 1–General. NIST Computer Security Resource Center*. Retrieved December 22, 2023, from https://csrc.nist.gov/pubs/sp/800/57/pt1/r5/final

Osborne, C. (2020, October 19). *This new malware uses remote overlay attacks to hijack your bank account*. ZDNET. Retrieved December 22, 2023, from https://www.zdnet.com/article/this-new-malware-uses-remote-overlay-attacks-to-hijack-your-bank-account/

Jerry Huang has worked as a technical and security staff at several prominent technology companies, gaining experience in areas like security, AI/ML, and large-scale infrastructure. At Metabase, an open-source business intelligence platform, he contributed features such as private key management and authentication solutions. As a Software Engineer at Glean, a Generative AI search startup, Jerry was one of the three engineers responsible for large-scale GCP infrastructure powering text summarization, autocomplete, and search for over 100,000 enterprise users. Previously at TikTok, Jerry worked to design and build custom RPCs to model access control policies. And at Roblox, he was a Machine Learning/Software Engineering Intern focused on real-time text generation models. He gathered and cleaned a large multilingual corpus that significantly boosted model robustness. Jerry has also conducted extensive security and biometrics research as a Research Assistant at Georgia Tech's Institute for Information Security & Privacy. This resulted in a thesis on privacy-preserving biometric authentication. His academic background includes a BS/MS in Computer Science from Georgia Tech and he is currently pursuing an MS in Applied Mathematics at the University of Chicago. phone: 571–268-6923;

Ken Huang is the author and chief editor of eight books on Generative Artificial Intelligence and Web3, published, respectively, by international publishers including Springer, Cambridge University Press, John Wiley, and China Machine Press. He currently serves as the CEO of the AI and Web3 consulting and education company DistributedApps.AI, based in the United States. Additionally, he holds multiple roles including the expert member of the Blockchain Committee of the Chinese Institute of Electronics, the Co-Chair of AI Organization Responsibility Working Group at Cloud Security Alliance, and Chair of the Blockchain Security Working Group at the Cloud Security Alliance, GCR. He is also a core contributor to the Generative AI Working Group at the NIST and a core author of the OWASP Top 10 for LLM Applications.

Ken Huang has been invited to provide Speaking or Consulting services at institutions including the University of California, Berkeley, Stanford University, Peking University, Tsinghua University, Shanghai Jiao Tong University, China Pacific Insurance, and the World Bank in the past.

Moreover, he has given keynote speeches at international conferences, such as:
– The Davos World Economic Forum 2020 Blockchain Conference
– Consensus 2018 in New York
– The American ACM AI & Blockchain Decentralized Annual Conference 2019
– IEEE Technology and Engineering Management Society Annual Meeting 2019
– Silicon Valley World Digital Currency Forum
– Sino-US Blockchain Summit in Silicon Valley

He has also been awarded the "Blockchain 60" Figure Award by the National University Artificial Intelligence and Big Data Innovation Alliance Blockchain Special Committee in China in 2021.

Winston Ma, CFA is an investor, author, and adjunct professor in the digital economy. He is one of a small number of native Chinese who has worked as investment professionals and practicing capital markets attorneys in both the United States and China. Most recently for 10 years, he was Managing Director and Head of the North America Office for China Investment Corporation (CIC), China's sovereign wealth fund.

Prior to that, Mr. Ma served as the deputy head of equity capital markets at Barclays Capital, a vice president at J.P. Morgan investment banking, and a corporate lawyer at Davis Polk & Wardwell LLP in New York.

At CIC's inception in 2007, he was among the first group of overseas hires by CIC, where he was a founding member of both CIC's Private Equity Department and later the Special Investment Department for direct investing (Head of CIC North America office 2014–2015). He had leadership roles in global investments involving financial services, technology (TMT), energy, and natural resources sectors, including the setup of West Summit (Huashan) Capital, a cross-border growth capital fund in Silicon Valley, which was CIC's first overseas tech investment. For global investments, he served on the board of internationally listed and private companies.

A nationally certified Software Programmer as early as 1994, Mr. Ma is the book author of China's Mobile Economy (Wiley 2016, among "best 2016 business books for CIOs"), Digital Economy 2.0 (2017 Chinese), The Digital Silk Road (2018 German), China's AI Big Bang (2019 Japanese), and Investing in China (Risk Books, 2006). His new books are "The Hunt for Unicorns: How Sovereign Funds Are Reshaping Investment in the Digital Economy" (Wiley October 2020) and "The Digital War—How China's Tech Power Shapes the Future of AI, Blockchain, and Cyberspace" (Wiley January 2021). He was selected as a 2013 Young Global Leader at the World Economic Forum (WEF) and has been a member of the Council for Long-Term Investing and the Council for Digital Economy and Society. He has been a member of the New York University (NYU) President's Global Council since its inception, and in 2014 he received the NYU Distinguished Alumni Award.

The manufacturer's authorised representative in the EU is Springer
Nature Customer Service Centre GmbH, Europaplatz 3, 69115 Heidelberg,
Germany. If you have any concerns regarding our products, please
contact ProductSafety@springernature.com

Printed and bound by CPI Group (UK) Ltd, Croydon, CR0 4YY

24/04/2026

02096316-0008